BIOMAGNETIC MONITORING OF PARTICULATE MATTER

BIOMAGNETIC MONITORING OF PARTICULATE MATTER

In the Indo-Burma Hotspot Region

Prabhat K. Rai
Department of Environmental Science
School of Earth Science and Natural Resource Management
Mizoram University, Aizawl, India

ELSEVIER

Amsterdam • Boston • Heidelberg • London • New York • Oxford
Paris • San Diego • San Francisco • Singapore • Sydney • Tokyo

Elsevier
Radarweg 29, PO Box 211, 1000 AE Amsterdam, Netherlands
The Boulevard, Langford Lane, Kidlington, Oxford OX5 1GB, UK
225 Wyman Street, Waltham, MA 02451, USA

Notices
Knowledge and best practice in this field are constantly changing. As new research and
experience broaden our understanding, changes in research methods, professional practices,
or medical treatment may become necessary.

Practitioners and researchers must always rely on their own experience and knowledge
in evaluating and using any information, methods, compounds, or experiments described
herein. In using such information or methods they should be mindful of their own safety
and the safety of others, including parties for whom they have a professional responsibility.

To the fullest extent of the law, neither the Publisher nor the authors, contributors, or
editors, assume any liability for any injury and/or damage to persons or property as a
matter of products liability, negligence or otherwise, or from any use or operation of any
methods, products, instructions, or ideas contained in the material herein.

ISBN: 978-0-12-805135-1

British Library Cataloguing-in-Publication Data
A catalogue record for this book is available from the British Library

Library of Congress Cataloging-in-Publication Data
A catalog record for this book is available from the Library of Congress

For information on all Elsevier publications
visit our website at http://store.elsevier.com/

Working together
to grow libraries in
developing countries

www.elsevier.com • www.bookaid.org

In the Loving Memory of my Grand Parents-Late Shri Braj Bihari Rai, Shrimati Chandravati Rai, Shri Rajkumar Rai (Ex Member of Parliament of India and cabinet minister Govt. of Uttar Pradesh, India) & my late son Ansh.

CONTENTS

ACKNOWLEDGMENTS

I consider it a rare opportunity to thank Professor A.N. Rai, former Director of NAAC and former Vice Chancellor of Mizoram University as well as North Eastern Hill University, for his blessings, encouragement, and support. I am also indebted for the love and affection of Mrs Urmila Rai for extending his love and blessings.

I am thankful to Professor Lalthantluanga, Vice Chancellor, Mizoram University for guidance and support. I am extremely thankful to Professor R.P. Tiwari, Department of Geology, Mizoram University (Vice Chancellor Hari Singh Gaur University, Sagar) who always helped and encouraged me. For academic guidance, I am thankful to Professor B.A. Maher (and her team, Professor Diwakar Tiwari, Professor B.P. Nautiyal, and Professor V.P. Khanduriz). I am also thankful to my teachers in Banaras Hindu University (BHU), especially Professor B.D. Tripathi, Professor J.S. Singh (FNA), Professor K.P. Singh, Professor L.C. Rai (FNA), Professor Rajeev Raman (FNA), Professor J.P. Gaur, Professor M. Agrawal, and Professor R.S. Upadhyay.

Many thanks to my research team namely Biku Moni Chutia (for collection of data), Pallab Deb, Sagosem Priyokumar Singh, and Muni Singh for their support. I would like to thank my friends including Alok Chourasia, Divesh Nath Sharma, Ramchandra, Nimesh Rai, and Mukesh Rai for always being with me.

I am also grateful to Respected Laura (Hon'ble Editor, Elsevier), Emily (editorial manager) and four anonymous reviewers for recommending this book and making critical suggestions.

I would like to thank my father, Dr Om Prakash Rai (principal of CHS-Banaras Hindu University); my mother, Usha Rai; brother, Prashant Rai; sister, Pratibha Rai; and my wife, Garima Rai, who have supported and encouraged me throughout this journey. Further, I am thankful to my brother-in-law Dr Ved Prakash Rai of Genetics and Plant Breeding, BHU (currently a scientist at Navsari University, Gujrat, India) for all his affection and encouragement. Moreover, I would like to extend my love to little Pranjali and Rachit as they brought a great deal of fortune with them.

Prabhat Kumar Rai

Prabhat Kumar Rai

Department of Environmental Sciences
School of Earth Sciences and Natural Resource Management
Mizoram Central University, Tanhril, Aizawl, 796004, India

ACKNOWLEDGMENT FOR FINANCE/ FUNDING/ACADEMIC GUIDANCE

The author is grateful to the Department of Biotechnology and Department of Science and Technology, Mizoram University for providing financial assistance in the form of this research project (vide project no. BT/PR-11889/BCE/08/730/2009 and SR/FTP/ES-83/2009, respectively). Thanks are due to Dr Onkar Nath Tiwari and Dr Umesh Sharma for their useful discussions and cooperation in this work. I am particularly grateful to Professor H.B. Singh, US Executive Editor in Chief, Atmospheric Environment, for giving shape to present the book in its current format. I am particularly thankful to Professor Cecil C Konijnendijk van den Bosch, Editor in Chief, Urban Forestry and Urban Greening, for his guidance.

CHAPTER ONE

Particulate Matter and Its Size Fractionation

1.1 INTRODUCTION

In the current Anthropocene, environmental pollution is a global problem that is inextricably linked with rapid industrialization and urbanization. Pollution hampers the environment sustainability and ecosystem services. In this chapter, we will briefly introduce environmental pollution (now popularly called pollution science) before the introduction of the theme topic.

1.1.1 Pollution Science

Environmental pollution is the unfavorable alteration of our surroundings, wholly or largely as a byproduct of man's actions, through direct or indirect effects of the changes in the energy pattern, radiation levels, and chemical and physical constitution and abundance of organisms. Environmental pollution is a global problem and is common to both developed as well as developing countries, which attracts the attention of human beings for its severe long-term consequences. The decline in environmental quality as a consequence of pollution is evidenced by loss of vegetation, biological diversity, excessive amounts of harmful chemicals in the ambient atmosphere and in food grains, and growing risks of environmental accidents and threats to life support systems. Pollution is viewed from different angles by different people but is commonly agreed to be the outcome of urban-industrial and technological revolution and rapacious and speedy exploitation of natural resources, increased rate of exchange of matter and energy, and ever-increasing industrial wastes, urban effluents, and consumer goods. Holdgate (1979) defined environmental pollution as the introduction by man, into the environment, of substances or energy liable to cause interference with legitimate uses of environment. Singh (1991) has defined pollution in a very simple manner, i.e., "Disequilibrium condition from equilibrium condition in any system." This definition may be applied to all

Biomagnetic Monitoring of Particulate Matter
ISBN 978-0-12-805135-1
http://dx.doi.org/10.1016/B978-0-12-805135-1.00001-9

types of pollution ranging from physical to economic, political, social, and religious. Over the past couple of decades, various sources of pollution were identified that altered the composition of water, air, and soil of the environment. The substances that cause pollution are known as pollutants. A pollutant can be any chemical (toxic metal, radionuclides, organophosphorus compounds, gases) or geochemical substance (dust, sediment), biological organism or product, or physical substance (heat, radiation, sound wave) that is released intentionally or inadvertently by man into the environment with actual or potential adverse, harmful, unpleasant, or inconvenient effects. Such undesirable effects may be direct (affecting man) or indirect, being mediated via resource organisms or climate change. Depending on the nature of pollutants and also subsequent pollution of environmental components, the pollution may be categorized as follows:

1. Air Pollution
2. Water Pollution
3. Soil/Land Pollution
4. Noise Pollution
5. Radioactive Pollution
6. Thermal Pollution

Among these types of pollution, air pollution is the main type threatening the environment, humans, plants, animals, and all living organisms.

1.2 AIR POLLUTION

Clean air, pure water, and nutritious food are basic amenities of life, and air is the most important resource for the sustenance of life and other activities in the biosphere. However, the quality of air, water, and land is deteriorating continuously. Air, being the lifeline, should be protected from the evils of pollution, as its quality depletion beyond a threshold limit may lead to serious health hazards to both living beings and vegetation. Air quality is generally described as a combination of the physical and chemical characteristics that make air a healthful resource for humans, animals, and plants (Joshi and Bora, 2011). All organisms need clean air for their healthy growth and development. But this air, which is so precious for life, has become highly polluted; it is obviously the first and foremost susceptible component of our environment prone to pollution.

Air pollution is the introduction of chemicals, particulate matter (PM), or biological materials into the atmosphere that cause harm or discomfort to humans or other living organisms, or cause damage to the environment

(Agbaire and Esiefarienrhe, 2009; Assadi et al., 2011). The World Health Organization (WHO, 1992, 1996) defines air pollution as substances put into the air by activity of mankind in concentrations sufficient to cause harmful effects to health, property, or crop yield, or to interfere with the enjoyment of property and considers air pollution to be a major environmental health problem deserving high priority for action. According to the Indian Air Amendment Act, 1987, air pollution means any solid, liquid, or gaseous substances present in the atmosphere in such concentration that may tend to be injurious to human beings or other living creatures or plants or property or enjoyment. Air pollution can also be defined as any abnormal increase or decrease in the concentration of the normal components of atmosphere (Mudakavi, 2010; Bhattacharya et al., 2012). Air pollution is one of the greatest environment evils and has become an extremely serious problem for the modern world and is of prime concern in India. The occurrence of air pollution was not perceived as a major problem in most countries until the late 1950s and 1960s. Until then, it was usually seen as a local problem in urban and industrial areas. Only in recent years have air pollution and other atmospheric issues evolved as problems of regional and global importance. Rapid industrialization, unplanned urbanization, an alarming increase in the number of vehicles, and population growth are considered to be the major causes of increased dust/particulate air pollution levels (Odilora et al., 2006; Jayanthi and Krishnamoorthy, 2006; Mandal, 2006; Agbaire and Esiefarienrhe, 2009; Kuddus et al., 2011). In cities, the main source of air pollution is the combustion of fossil fuels from automobiles, diesel trucks, ships, and construction equipment (mobile sources), and from heating furnaces and power plants (stationary sources). A variety of industrial processes such as steel mills and cement kilns can also significantly contribute to air pollution, while in many parts of the developing world cooking fires continue to play a role. The specific composition of this mixture is variable across locations due to differences in geography, climate, and emission sources. Various physical, chemical, and dynamic processes may generate air pollutants including particulates and gaseous contaminants that may cause adverse health effects in humans or animals, affect plant life, and impact the global environment by changing the atmosphere of the earth (Raabe, 1999; Bakand et al., 2005; Hayes et al., 2007). Industrialization and automobiles are responsible for a maximum amount of air pollutants (Joshi and Swami, 2007). Climate conditions, the physicochemical properties of air pollutants, and how much time they linger in the atmosphere impact the surrounding plants and animals (Wagh et al., 2006). The places from which air pollutants emanate are called "sources" and the places where air pollutants disappear from the polluted air are called "sinks." Sinks include soil, vegetation, monuments, and water bodies (particularly oceans).

Air pollutants can be classified as either primary or secondary. Primary air pollutants are substances that are directly emitted into the atmosphere from natural and anthropogenic sources, whereas secondary pollutants are not emitted directly, but rather, they form in the air when primary pollutants react or interact in the atmosphere. The main air pollutants are represented by gaseous forms, particles in suspension, different ionizing radiation, and noise. The gaseous pollutants include oxidized and reduced forms of carbon, nitrogen, sulfur dioxide, ozone, carbon monoxide (CO), volatile organic compounds (VOCs), etc. The particulate forms include PM_{10} and $PM_{2.5}$ PM, heavy metals with toxic effect, polycyclic aromatic hydrocarbons, etc. The US Environmental Protection Agency (EPA) and ambient air standards define six major pollutants: carbon monoxide, lead, nitrogen dioxide, ozone, PM, and sulfur dioxide. These primary pollutants together contribute more than 90% of global air pollution. Particulate pollutants are the most dangerous among the primary pollutants (relative toxicity–particulate = 107 compared to CO = 1) (De, 2010). It is well known that air pollution, particularly particulates, originating from multiple sources (industries, vehicles, rocks, mines, and quarries), represents a threat to the environment as well as human health. Of the common air pollutants, PM has been studied in the greatest detail due to significant and increasing production by diverse vehicles. However, in this chapter we are confining our discussion to particulate pollution.

1.3 PARTICULATE MATTER

Particulate matter is a mixture of particles and droplets in the air, consisting of a variety of components such as organic compounds, metals, acids, soil, and dust (US Environmental Protection Agency, 1996; Ciencewicki and Jaspers, 2007). PM has been widely studied in recent years and the United Nations estimated that in the early twenty-first century over 600 million people in urban areas worldwide were exposed to dangerous levels of traffic-generated air pollutants (Cacciola et al., 2002).

PM is either directly emitted into the atmosphere from various natural and anthropogenic sources or can be formed from gases through chemical reactions. PM air pollution is derived from vehicle emissions, forest fires, and industrial, domestic, and agricultural pollutants. A wide range of natural and anthropogenic emission sources contribute to PM concentrations in the atmosphere such as windblown soil dust, marine and biogenic aerosols, road traffic and off-road vehicles, stationary combustion processes, industrial and construction processes, and combustion of agricultural waste (El-Fadel and Massoud, 2000). Fine particles are characterized by their etiology, their ability to remain suspended in the air and to carry material that is absorbed on the surface. The smaller the particle diameter, the longer it remains suspended in the air and the more hazardous it is.

Atmospheric PM with aerodynamic diameter <10 mm (PM_{10}) or <2.5 m ($PM_{2.5}$) are of considerable concern for public health (Schwartz et al., 1996a; NEPC, 1998; Beckett et al., 1998; Borja-Aburto et al., 1998; Pope III, 2000; Prajapati and Tripathi, 2008a,b; Rai, 2013). The ultrafine particles with typical dimension of nanometer-length scale are the most hazardous (Wahlin et al., 2006) as they cause several life-threatening diseases (Samet et al., 2000; Veranth et al., 2003; Brook et al., 2004; Wahlin et al., 2006). Ultrafine particles are responsible for the bulk of adverse health effects associated with particles in ambient air (Penttinen et al., 2001; Rai, 2013) and are more potent than fine or coarse PM in their ability to induce cellular damage (Le et al., 2002). They also pass very rapidly into the circulatory system (Nemmar et al., 2001). Suspended particulate matter (SPM) is of the greatest concern as it contributes 50% to total air pollution and causes respiratory disorders in human beings upon prolonged exposure (Freer-Smith et al., 2004) as it include all airborne particles in the size range of 0.5–100 μm. Sirajuddin and Ravichandran (2010) also studied SPM-related respiratory disorders such as nose block, sneezing, coughing, and hyperacidity in Tiruchirappalli, India. Bhattacharjee et al. (2010), noted that PM_S < 4.6 μm and PM_S < 1.1 μm are hazardous to human health due to their capacity to be inhaled into the bronchial region and deposited in the alveolar region. Epidemiologic findings suggest that short-term PM exposure can trigger acute or terminal health events whereas long term PM exposure can promote life-shortening chronic illness. Additional evidence suggests that particulate matter exposure over time can alter lung function, lung tissue and structure, airway responsiveness and respiratory defense mechanisms, and can increase susceptibility to respiratory infection and damage respiratory cells (EPA, 1996, 1997).

Considering that more than half of the global population now lives in urban settings (WHO, 2011; Hofman et al., 2014), a large exposure to urban air pollution can be expected. Currently, more than 85% of the European Union's (EU's) urban population is exposed to atmospheric particulate matter levels above the 2005 WHO Air Quality Guidelines (EEA, 2013; Hofman et al., 2014).

A second major concern is the ability of airborne particles to impact climate through absorption or scattering of solar radiation (Charlson et al., 1992; Haywood and Shine, 1995; Schwartz, 1996b), alteration of cloud properties (Charlson et al., 1992; Jones et al., 1994; Boucher and Lohmann, 1995; Haywood and Boucher, 2000), and decreasing surface albedo after deposition to snow and ice (Hansen and Nazarenko, 2004; Jacobson, 2004; Roberts and Jones, 2004). With an atmospheric residence time ranging from days to weeks, particulate matter is not only a local concern but a global one, the generation of pollutants in one region impacting the air quality of another. There may be several kinds of airborne particulates such as dust, smoke, fumes, mist, fog, smog, haze, etc.

Dust causes some of the highest concentrations of ambient primary PM in many areas around the globe. Environmental contamination and human exposure with respect to dust pollution have dramatically increased during the past 10 years (Faiz et al., 2009). Solid matter, which is composed of soil, anthropogenic metallic constituents, and natural biogenic materials, is called dust/particulate (Ferreira-Baptista and DeMiguel, 2005; Fathi and Clare, 2011). The particles of dust that deposit from the atmosphere and accumulate along roadsides are called road dust particles and originate from the interaction of solid, liquid, and gaseous metals (Akhter and Madany, 1993; Faiz et al., 2009; Fathi and Clare, 2011). It is estimated that dust pollution comprises around 40% of the total air pollution problem in India (Khan et al., 1989).

Dust pollution in the atmosphere, particularly of PM_{10}, is of current concern worldwide due to adverse health effects associated with their inhalation (Morris et al., 1995; Oberdorster, 2000; Pope et al., 2004; Calderon-Garciduenas et al., 2004; Faiz et al., 2009; Rai, 2011a,b; 2013). Moreover, PM in dust is thought to be the most harmful pollution component widely present in the environment (Bealey et al., 2007; Rai, 2013). Furthermore, the implication of the intake of dust particles with high concentration of heavy metals is that it poses potentially deleterious effects on the health of human beings (Faiz et al., 2009; Fathi and Clare, 2011).

Particulate matter is closely associated with increases in morbidity and mortality (Maier et al., 2008; Abbas et al., 2009). The London smog of December 1952 is estimated to have caused some 4000 excess deaths (Harrison and Yin, 2000). In the world's current phase of rapid industrialization and urbanization, the problem of PM pollution is a serious health concern for both developing and developed countries.

Solid matter, which is composed of soil, anthropogenic metallic constituents, and natural biogenic materials, is called dust (Ferreira-Baptista and DeMiguel, 2005; Rai, 2013). PM refers to a suspension of solid, liquid, or a combination of solid and liquid particles in the air (Hinds, 1999; Wilson et al., 2005). Air pollution originating from PM is generally characterized by its highly complex nature (Alfaro-Moreno et al., 2002; Abbas et al., 2009). PM is a mixture of particles and droplets in the air, consisting of a variety of components such as organic compounds, metals, acids, soil, and dust (Ciencewicki and Jaspers, 2007; Rai et al., 2014; Rai and Panda, 2014).

PM is one of six "criteria pollutants" designated by the US Clean Air Act of 1971 (Wilson et al., 2005). PM belongs to the class of poorly soluble particles that also encompass carbon black, coal mine dust, and titanium dioxide (Borm et al., 2004; Moller et al., 2008). Measurements of PM in ambient air are usually reported as the mass of particles with an aerodynamic diameter that is less than 2.5 μm or 10 μm (Zhu et al., 2006). These particle sizes are emphasized in view of their pertinent health impacts (Jahn et al., 2011; Rohr and Wyzga, 2012; Taner et al., 2013; Rai, 2013; Hicken et al., 2014; Pascal et al., 2014; Yadav et al., 2014; Rai and Panda, 2014; Rai et al., 2014; Kim et al., 2015; Yang et al., 2015).

Urban air quality is continually affected by emissions from stationary and mobile combustion sources. Mobile sources contribute to the emission of major urban air pollutants including carbon dioxide (CO_2), nitrogen oxides (NO_x), sulfur oxides (SO_x), PM, lead (Pb), photochemical oxidants, such as ozone (O_3), and ozone precursors like hydrocarbons and volatile organic compounds (Costa, 2004). Among these pollutants, the concentration of PM_{10} and $PM_{2.5}$ airborne aerosols have shown good agreement with traffic-released pollutants and other combustion processes (Prajapati and Tripathi, 2007).

1.4 SOURCES OF PM POLLUTION

Sources of PM pollution may be natural as well as anthropogenic. Natural processes that emit PM into the atmosphere include volcanic eruptions, geochemical sources, windblown dust, soil, and spray from marine sources. Natural sources of PM, e.g., volcanic eruptions, may contain sulfurous particles and may have an adverse effect on cardiorespiratory health in adults (Longo et al., 2008). Anthropogenic sources include power plants, traffic, agriculture, and various industrial activities such as mining and the metallurgical industries. Diesel exhaust emissions are the major source of $PM_{2.5}$ in the urban environment. Vehicular emissions consist of PM and gaseous emissions, with biologically active carbonaceous products present in both phases. Black carbon, mainly from diesel fuels, is found in ultrafine and fine-sized fractions, mainly smaller than 1 μm and predominantly smaller than 0.18 μm. Such vehicular particulates are often coated with condensed organic and inorganic compounds. Ambient concentrations of the platinum group elements (PGEs) platinum (Pt), palladium (Pd), and rhodium (Rh) have been on the rise, largely due to the use of automobile catalytic converters, which employ these metals as exhaust catalysts; these PGEs may impose a considerable human health risk (Wiseman and Zereini, 2009). Approximately 75% of diesel $PM_{2.5}$ emissions consist of such carbons. Results of a case study of restaurants in Turkey indicated that cooking is a significant source of indoor particulate matter that can cause adverse health effects; the lifetime cancer risk associated with As and Cr (VI) exposure was significant at selected restaurants, which might be of concern for restaurant workers (Taner et al., 2013).

Anthropogenic emissions in conjunction with the topographical and meteorological conditions can result in high air pollution within cities, which may inhibit human performance levels, as predicted before the Athens Olympics. In Udaipur region, Rajasthan, India, PM increase may be associated with high morbidity, particularly during the winter season (Yadav et al., 2014).

Theophanides et al. (2007) evaluated the atmospheric pollution created by industry and traffic areas near to the Greek cities, and the corresponding mortality of citizens in the region, and found that adverse environmental impact of air pollutants is a major concern in the industrial centers.

1.5 SIZE FRACTIONATION OF PM

Particulate matter can be classified as coarse, fine, and ultrafine depending upon the particle size. PM measured in urban air used in health effects studies and for regulation includes:

- Nuclei mode (smaller than 0.1 μm), often referred to as ultrafine particles; they do not last long in the air since they deposit or rapidly form fine particles by coagulation.
- Accumulation mode (between 0.1 μm and approximately 1.0–2.5 μm) account for the majority of the mass of suspended particles and deposit slowly leading to a long atmospheric lifetime of 5–10 days and the

buildup of visible haze. These particles may readily penetrate indoor spaces and are most strongly linked to adverse health effects.

- Coarse mode (larger than 1 μm), which extends up to 100 μm; they deposit relatively quickly with a lifetime of less than 2 days (Robert et al., 2003).

Dust pollution in the atmosphere, particularly of pollutant particles below 10 μm, is of current concern worldwide due to adverse health effects associated with their inhalation (Morris et al., 1995; Oberdorster, 2000; Pope et al., 2004; Calderon-Garciduenas et al., 2004; Faiz et al., 2009; Rai, 2011a,b, 2013). Moreover, PM in dust is thought to be the most harmful pollution component widely present in the environment (Bealey et al., 2007; Rai, 2013). In this chapter, we introduced the concept of environmental pollution with special reference to PM pollution and its associated effects. In the next chapter, we will confine our discussion to the effects of PM pollution on human health.

REFERENCES

Abbas, et al., 2009. Air pollution particulate matter (PM$_{2.5}$)-induced gene expression of volatile organic compound and/or polycyclic aromatic hydrocarbon-metabolizing enzymes in an in vitro coculture lung model. Toxicology In Vitro 23, 37–46.

Agbaire, P.O., Esiefarienrhe, E., 2009. Air pollution tolerance indices (apti) of some plants around otorogun gas plant in Delta State, Nigeria. Journal of Applied Sciences and Environmental Management 13 (1), 11–14.

Akhter, M.S., Madany, I.M., 1993. Heavy metals in street and house dust in Bahrain. Water, Air and Soil Pollution 66, 111–119.

Alfaro-Moreno, E., Martinez, L., Garcia-Cuellar, C., Bonner, J.C., Murray, J.C., Rosas, I., Rosales, S.P., Miranda, J., Osornio-Vargas, A.R., 2002. Biological effects induced in vitro by PM$_{10}$ from three different zones of Mexico City. Environmental Health Perspectives 110, 715–720.

Assadi, A., Pirbalouti, A.G., Teimori, N., Assadi, L., 2011. Impact of air pollution on physiological and morphological characteristics of *Eucalyptus camaldulensis* Den. Journal of Food, Agriculture and Environment 9 (2), 676–679.

Bakand, S., Winder, C., Khalil, C., Hayes, A., 2005. Toxicity assessment of industrial chemicals and airborne contaminants: transition from in vivo to in vitro test method: a review. Inhalation Toxicology 17 (13), 775–787.

Bealey, W.J., Mcdonald, A.G., Nemitz, E., Donovan, R., Dragosits, U., Duffy, T.R., Fowler, D., 2007. Estimating the reduction of urban PM$_{10}$ concentrations by trees within an environmental information system for planners. Journal of Environmental Management 85, 44–58.

Beckett, K.P., Freer-Smith, P.H., Taylor, G., 1998. Urban woodlands: their role in reducing the effects of particulate. Environmental Pollution 99, 347–360.

Bhattacharjee, A., Mandal, R., Chini, K.T., 2010. A preliminary study on the nature of particulate matters in vehicle fuel wastes. Environmental Monitoring and Assessment. http://dx.doi.org/10.1007/S10661-010-1598-X.

Bhattacharya, T., Chakraborty, S., Kagathara, M., Thakur, B., 2012. Ambient air quality and the air pollution tolerance indices of some common plant species of Anand city, Gujarat, India. Report and Opinion 4 (9), 7–15.

Borja-Aburto,V.H., Castillejos, M., Gold, D.R., Bierzwinski, S., Loomis, D., 1998. Mortality and ambient fine particles in southwest Mexico city, 1993–1995. Environmental Health Perspectives 106, 849–855.

Borm, P.J., Schins, R.P., Albrecht, C., 2004. Inhaled particles and lung cancer, part B: paradigms and risk assessment. International Jounal of Cancer 110, 3–14.

Boucher, O., Lohmann, U., 1995. The sulfate-CCN-cloud albedo effect—a sensitivity study with 2 general-circulation models. Tellus Series B-Chemical and Physical Meteorology 47 (3), 281–300.

Brook, R.D., Franklin, B., Cascio, W., Hong, Y.L., Howard, G., Lipsett, M., et al., 2004. Air pollution and cardiovascular diseasea statement for healthcare professionals from the expert panel on population and prevention science of the American Heart Association. Circulation 109 (21), 2655–2671.

Cacciola, R.R., Sarva, M., Polosa, R., 2002. Adverse respiratory effects and allergic susceptibility in relation to particulate air pollution: flirting with disaster. Allergy 57, 281–286.

Calderon-Garciduenas, L., Reed, W., Maronpot, R.R., Henriquez-Roldan, C., Delago-Chavez, R., Calderon-Garciduenas, A., Dragustinovis, I., Franco-Lira, M., Aragon-Flores, M., Solt, A.C., Altenburg, M., Torres-Jardon, R., Swenburg, J.A., 2004. Brain inflammation and Alzheimer's-like pathology in individuals exposed to severe air pollution. Toxicologic Pathology 32, 650–658.

Charlson, R.J., et al., 1992. Climate forcing by anthropogenic aerosols. Science 255 (5043), 423–430.

Ciencewicki, J., Jaspers, I., 2007. Air pollution and respiratory viral infection. Inhalation Toxicology 19 (14), 1135–1146.

Costa, D.L., 2004. Issues that must be addressed for risk assessment of mixed exposures: the US EPA experience with air quality. Journal of Toxicology and Environmental Health, Part A 67 (3), 195–207.

De, A.K., 2010. Text Book of Environmental Chemistry, seventh ed. New Age International Limited.

EEA, 2013. Air Quality in Europe e 2013 Report. European Environment Agency, Copenhagen, Denmark.

El-Fadel, M., Massoud, M., 2000. Particulate matter in urban areas: health-based economic assessment. The Science of the Total Environment 257, 133–146.

Environmental Protection Agency, (U.S.), 1996. Air Quality Criteria for Particulate Matter. Office of Research and Development, Office of Health and Environmental Assessment, Research Triangle Park, N.C. EPA report no. EPA/600/P-95/001aF.

Environmental Protection Agency (E.P.A.-US), 1997. Regulatory Impact Analysis for the Particulate Matter and Ozone. National Ambient Air Quality Standards and Proposed Regional Haze Rule. Research Triangle Park, N.C. Innovative Strategies and Economics Group, Office of Air Quality Planning and Standards, USA.

Faiz, Y., Tufail, M., Javed, M.T., Chaudhry, M.M., Siddique, N., 2009. Road dust pollution of Cd, Cu, Ni, Pb and Zn along Islamabad Expressway, Pakistan. Micro Chemical Journal 92, 186–192.

Fathi, Z., Clare, W., 2011. Urban Airborne Particulate Matter- Origin, Chemistry, Fate and Health Impacts, XXIII. Springer Verlag. 656 p.

Ferreira-Baptista, L., DeMiguel, E., 2005. Geochemistry and risk assessment of street dust in Luanda, Angola: a tropical urban environment. Atmospheric Environment 39, 4501–4512.

Freer-Smith, P.H., El-Khatib, A.A., Taylor, G., 2004. Capture of particulate pollution by trees: a comparison of species typical of semi-arid areas (*Ficus nitida* and *Eucalyptus globulus*) with European and North-American species. Water, Air, and Soil Pollution 155, 173–187.

Hansen, J., Nazarenko, L., 2004. Soot climate forcing via snow and ice albedos. Proceedings of the National Academy of Sciences of the United States of America 101 (2), 423–428.

Harrison, R.M., Yin, J., 2000. Particulate matter in the atmosphere: which particle properties are important for its effects on health? The Science of the Total Environment 249, 85–101.

Hayes, A., Bakand, S., Winder, C., 2007. Novel in vitro exposure techniques for toxicity testing and biomonitoring of airborne contaminants. In: Drug Testing In Vitro-Achievements and Trends in Cell Culture Techniques. Wiley-VCH, Berlin, pp. 103–124.

Haywood, J.M., Shine, K.P., 1995. The effect of anthropogenic sulfate and soot on the clear sky planetaryradiation budget. Geophysical Research Letters 22, 603–606.

Haywood, J., Boucher, O., 2000. Estimates of the direct and indirect radiative forcing due to tropospheric aerosols: a review. Reviews of Geophysics 38 (4), 513–543.

Hicken, M.T., et al., 2014. Fine particulate matter air pollution and blood pressure: the modifying role of psychosocial stress. Environmental Research 133, 195–203.

Hinds, W.C., 1999. Aerosol Technology: Properties, Behavior, and Measurements of Airborne Particles. Wiley, New York.

Hofman, J., Wuyts, K., Wittenberghe, S.V., Brackx, M., Samson, R., 2014. On the link between biomagnetic monitoring and leaf-deposited dust load of urban trees: relationships and spatial variability of different particle size fractions. Environmental Pollution 189, 63–72.

Holdgate, M.W., 1979. A Perspective of Environmental Pollution. Cambridge Univ. Press, Cambridge.

Jacobson, M.Z., 2004. Climate response of fossil fuel and biofuel soot, accounting for soot's feedback to snow and sea ice albedo and emissivity. Journal of Geophysical Research-Atmospheres 109 (D21).

Jahn, H.J., et al., 2011. Particulate matter pollution in the megacities of the Pearl River Delta, China—a systematic literature review and health risk assessment. International Journal of Hygiene and Environmental Health 214 (4), 281–295.

Jayanthi, V., Krishnamoorthy, R., 2006. Status of ambient air quality at selected sites in Chennai. International Journal of Environment and Pollution 25, 696–704.

Jones, A., Roberts, D.L., Slingo, A., 1994. A climate model study of indirect radiative forcing by anthropogenic sulfate aerosols. Nature 370 (6489), 450–453.

Joshi, N., Bora, M., 2011. Impact of air quality on physiological attributes of certain plants. Report and Opinion 3 (2), 42–47.

Joshi, P.C., Swami, A., 2007. Physiological responses of some tree species under roadside automobile pollution stress around city of Haridwar, India. Environmentalist 27, 365–374.

Khan, A.M., Pandey, V., Yunus, M., Ahmad, K.J., 1989. Plants as dust scavengers - a case study. The Indian Forester 115 (9), 670–672.

Kim, K., Kabir, E., Kabir, S., 2015. A review on the human health impact of airborne particulate matter. Environment International 74, 136–143.

Kuddus, M., Kumari, R., Ramteke, W.P., 2011. Studies on air pollution tolerance of selected plants in Allahabad city, India. Journal of Environmental Research and Management 2 (3), 042–046.

Le Tertre, A., Medina, S., Samoli, E., Forsberg, B., Michelozzi, P., Boumghar, A., et al., 2002. Short-term effects of particulate air pollution on cardiovascular diseases in eight European cities. Journal Epidemiology and Community Health 56, 773–779.

Longo, B.M., Rossignol, A., Green, J.B., 2008. Cardiorespiratory health effects associated with sulphurous volcanic air pollution. Public Health 122, 809–820.

Maier, K.L., Alessandrini, F., Beck-Speier, I., Hofer, T.P.J., Diabaté, S., Bitterle, E., Stöger, T., Jakob, T., Behrend, H., Horsch, M., Beckers, J., Ziesenis, A., Hültner, L., Frankenberger, M., Krauss-Etschmann, S., Schulz, H., 2008. Health effects of ambient particulate matter—biological mechanisms and inflammatory responses to in vitro and in vivo particle exposures. Inhalation Toxicology 20, 319–337.

Mandal, M., 2006. Physiological changes in certain test plants under automobile exhaust pollution. Journal of Environmental Biology 27 (1), 43–47.

Moller, P., et al., 2008. Air pollution, oxidative damage to DNA, and carcinogenesis. Cancer Letters 266, 84–97.

Morris, W.A., Versteeg, J.K., Bryant, D.W., Legzdins, A.E., Mccarry, B.E., Marvin, C.H., 1995. Preliminary comparisons between mutagenicity and magnetic susceptibility of respirable airborne particulate. Atmospheric Environment 29, 3441–3450.

Mudakavi, J.R., 2010. Principles of Practices of Air Pollution Control and Analysis. I.K. International Publishing house Pvt. Ltd, New Delhi, India, pp. 8–10.

Nemmar, A., Vanbilloen, H., Hoylaerts, M.F., Hoet, P.H., Verbruggen, A., Nemery, B., 2001. Passage of intratracheally instilled ultrafine particles from the lung into the systemic circulation in hamster. American Journal of Respiratory and Critical Care Medicine 164 (9), 1665–1668.

NEPC, 1998. Ambient Air Quality: National Environment Protection Measure for Ambient Air Quality. National Environment Protection Council Service Corporation, Adelaide.

Oberdorster, G., 2000. Toxicology of ultrafine particles: in vivo studies. Philosophical Transactions of the Royal Society of London A358, 2719–2740.

Odilora, C.A., Egwaikhide, P.A., Esekheigbe, A., Emua, S.A., 2006. Air pollution tolerance indices (APTI) of some plant species around llupeju Industrial area, Lagos. Journal of Engineering Science and Application 4 (2), 97–101.

Pascal, M., et al., 2014. Short-term impacts of particulate matter (PM_{10}, $PM_{10-2.5}$, $PM_{2.5}$) on mortality in nine French cities. Atmospheric Environment 95, 175–184.

Penttinen, P., Timonen, K.L., Tiittanen, P., Mirme, A., Ruuskanen, J., Pekkanen, J., 2001. Ultrafine particles in urban air and respiratory health among adult asthmatics. European Respiratory Journal 17, 428–435.

Pope, C.A., Hansen, M.L., Long, R.W., Nielsen, K.R., Eatough, N.L., Wilson, W.E., Eatough, D.J., 2004. Ambient particulate air pollution, heart rate variability, and blood markers of inflammation in a panel of elderly subjects. Environmental Health Perspectives 112, 339–345.

Pope III, C.A., 2000. Epidemiology of Fine particulate air pollution and human health: biologic mechanisms and who's at risk? Environmental Health Perspectives 108 (4), 713–723.

Prajapati, S.K., Tripathi, B.D., 2007. Biomonitoring trace-element levels in PM_{10} released from vehicles using leaves of *Saraca indica* and *Lantana camara*. AMBIO: A Journal of Human Environment 36 (8), 704–705.

Prajapati, S.K., Tripathi, B.D., 2008a. Anticipated performance index of some tree species considered for green belt development in and around an urban area: a case study of Varanasi City, India. Journal of Environmental Management 88 (4), 1343–1349.

Prajapati, S.K., Tripathi, B.D., 2008b. Biomonitoring seasonal variation of urban air polycyclic aromatic hydrocarbons (PAHs) using *Ficus benghalensis* leaves. Environmental Pollution 151, 543–548.

Raabe, O.G., 1999. Respiratory exposure to air pollutants. In: Swift, D.L., Foster, W.M. (Eds.), Air Pollutants and the Respiratory Tract. Marcel Dekker INC, NY, USA, pp. 39–73.

Rai, P.K., 2011a. Dust deposition capacity of certain roadside plants in Aizawl, Mizoram: implications for environmental geomagnetic studies. In: Dwivedi, S.B., et al. (Ed.), Recent Advances in Civil Engineering, pp. 66–73.

Rai, P.K., 2011b. Biomonitoring of particulates through magnetic properties of road-side plant leaves. In: Tiwari, D. (Ed.), Advances in Environmental Chemistry. Excel India Publishers, New Delhi, pp. 34–37.

Rai, P.K., 2013. Environmental magnetic studies of particulates with special reference to biomagnetic monitoring using roadside plant leaves. Atmospheric Environment 72, 113–129.

Rai, P.K., Panda, L.S., 2014. Dust capturing potential and air pollution tolerance index (APTI) of some road side tree vegetation in Aizawl, Mizoram, India: an Indo-Burma hot spot region. Air Quality, Atmospheric and Health 7 (1), 93–101.

Rai, P.K., Chutia, B.M., Patil, S.K., 2014. Monitoring of spatial variations of particulate matter (PM) pollution through bio-magnetic aspects of roadside plant leaves in an Indo-Burma hot spot region. Urban Forestry and Urban Greening 13, 761–770.

Robert, D., Brook, J., Brook, R., Sanjay, R., 2003. Air pollution: the "Heart" of the problem. Current Hypertension Reports Current Science 5, 32–39.

Roberts, D.L., Jones, A., 2004. Climate sensitivity to black carbon aerosol from fossil fuel combustion. Journal of Geophysical Research-Atmospheres 109 (D16).

Rohr, C.A., Wyzga, R.E., 2012. Attributing health effects to individual particulate matter constituents. Atmospheric Environment 62, 130–152.

Samet, J.M., Dominici, F., Curriero, F.C., Coursac, I., Zeger, S.L., 2000. Fine particulate air pollution and mortality in 20 U.S. cities, 1987–1994. New England Journal of Medicine 343 (24), 1742–1749.

Schwartz, J., 1996a. Air pollution and hospital admissions for respiratory disease. Epidemiology 7, 20–28.

Schwartz, S.E., 1996b. The Whitehouse effect–shortwave radiative forcing of climate by anthropogenic aerosols: an overview. Journal of Aerosol Science 27 (3), 359–382.

Singh, S., 1991. Environmental Geography. Prayag Pustak Bhawan, pp. 466–507.

Sirajuddin, M., Ravichandran, M., 2010. Ambient air quality in an urban area and its effects on plants and human beings: a case study of Tiruchiraalli, India. Kathmandu University Journal of Science, Engineering and Technology 6 (2), 13–19.

Taner, S., Pekey, B., Pekey, H., 2013. Fine particulate matter in the indoor air of barbeque restaurants: elemental compositions, sources and health risks. Science of the Total Environment 454-455, 79–87.

Theophanides, M., Anastassopoulou, J., Vasilakos, C., Maggos, T., Theophanides, T., 2007. Mortality and pollution in several greek cities. Journal of Environmental Science and Health, Part A 42 (6), 741–746.

Veranth, J.M., Gelein, R., Oberdorster, G., 2003. Vaporization-condensation generation of ultrafine hydrocarbon particulate matter for inhalation toxicology studies. Aerosol Science and Technology 37, 603–609.

Wagh, N.D., Shukla, P.V., Tambe, S.B., Ingle, S.T., 2006. Biological monitoring of roadside plants exposed to vehicular pollution in Jalgaon city. Journal of Environmental Biology 27 (2), 419–421.

Wahlin, P., Berkowicz, R., Palmagren, F., 2006. Characterisation of traffic-generated particulate matter in Copenhagen. Atmospheric Environment 40, 2151–2159.

WHO, 2011. Fact Sheet: Air Quality and Health. WHO Media Centre.

Wilson, J.G., Kingham, S., Pearce, J., Sturman, A.P., 2005. A review of intraurban variations in particulate air pollution: implications for epidemiological research. Atmospheric Environment 39, 6444–6462.

Wiseman, C.L.S., Zereini, F., 2009. Airborne particulate matter, platinum group elements and human health: a review of recent evidence. Science of the Total Environment 407, 2493–2500.

World Health Organization, 1992. Our Planet Our Health: Report of the WHO Commission on Health and Environment. Geneva.

World Health Organization, 1996. Regional Health Report. Regional office for south East Asia, New Delhi.

Yadav, R., Beig, G., Jaffrey, S.N.A., 2014. The linkages of anthropogenic emissions and meteorology in the rapid increase of particulate matter at a foothill city in the Arawali range of India. Atmospheric Environment 85, 147–151.

Yang, T.H., et al., 2015. Personal exposure to particulate matter and inflammation among patients with periodontal disease. Science of the Total Environment 502, 585–589.

Zhu, Y., Kuhn, T., Mayo, P., Hinds, W.C., 2006. Comparison of daytime and nighttime concentration profiles and size distributions of ultrafine particles near a major highway. Environmental Science and Technology 40, 2531–2536.

CHAPTER TWO

Adverse Health Impacts of Particulate Matter

2.1 INTRODUCTION

Urban air quality is becoming a serious public health concern at the global scale. As discussed in the previous chapter, particulate matter (PM) pollution is intimately linked with human health. This chapter describes the different human health implications associated with PM pollution. As introduced in the first chapter, PM may derive its origin from natural and anthropogenic sources. Vehicle-derived pollutants as well as industrial emissions simultaneously release deleterious fine-grained PM into the atmosphere. Fine PM, especially 2.5 μm ($PM_{2.5}$) and 10 μm (PM_{10}) are particularly harmful to human health. Air pollution in the form of PM is an important environmental health risk factor for respiratory and cardiovascular morbidity and mortality. Further, PM is inextricably linked with genotoxicity and mutations. A literature review of the cellular and molecular basis of adverse effects associated with PM is presented in this chapter.

2.2 PARTICULATE MATTER (PM) AND HEALTH

Particulate matter is associated with many adverse human health impacts (Pope, 2000; Jahn et al., 2011; Rohr and Wyzga, 2012; Taner et al., 2013; Raaschou-Nielsen et al., 2013; Beelen et al., 2013; Hamra et al., 2014; Pascal et al., 2014; Rai and Panda, 2014; Rai et al., 2014; Dergham et al., 2015; Hicken et al., 2014; Yadav et al., 2014; Kim et al., 2015; Yang et al., 2015). PM vehicular emissions, notably in the ultrafine fraction, have been specifically associated with endpoints such as oxidative stress and mitochondrial damage (Li et al., 2003), lipid peroxidation (Pereira et al., 2007), upregulation of genes relevant to vascular inflammation (Gong et al., 2007), and early atherosclerosis and oxidative stress (Araujo et al., 2008). Progression of atherosclerosis has also been reported due to exposure of PM pollution (Suwa et al., 2002). Table 2.1 lists the global research on adverse health impacts of PM pollution.

Biomagnetic Monitoring of Particulate Matter
ISBN 978-0-12-805135-1
http://dx.doi.org/10.1016/B978-0-12-805135-1.00002-0

Table 2.1 List of adverse impacts of PM demonstrated through global research

Effects of particulate pollutants on human health

Particulate matter	Health impacts	References
Ultrafine particles and fine particle suspended particulate matter (SPM)	Ultrafine particles induce vascular and systemic inflammation, oxidative stress, cellular damage, mitochondrial damage, lipid per oxidation, and early atherosclerosis. Fine particles may provoke alveolar inflammation, resulting in the release of harmful cytokines and increased blood coagulability. Fine PM may cause mild eye irritation, mortality, and respiratory disorders such as nose block, sneezing, cough, and hyperacidity. It also affects infant birth weight and mortality, causing sudden infant death syndrome. Ambient particulate matter exposure can be associated with specific physiologic endpoints including reduced lung function causing lung inflammation and injury-increased blood plasma viscosity affecting vascular tone and endothelial function, reduced heart rate variability, increased circulating markers of inflammation, mild hypoxemia, or decline in blood oxygen saturation. Increased symptoms of obstructive airway disease include chronic cough, bronchitis, and chest illness. PM is also associated with decreases in DNA methylation in *NOS24*, a gene directly responsible for production of nitric oxide, an important player in both respiratory and cardiovascular diseases. Elevated concentration of $PM_{2.5}$ exposure is strongly associated with myocardial infarction, ischemic heart disease, dysrhythmias, heart failure, cardiac arrest, increased carotid intima-media thickness, a measure of subclinical atherosclerosis. Short-term exposure to diesel exhaust mainly consisting of $PM_{2.5}$ has an acute inflammatory effect resulting in marked neutrophilia, activation of mast cells and neutrophils and the production of cytokines and chemokines associated with neutrophil accumulation and activation.	Seaton et al. (1995), Driscoll et al. (1997), Donaldson et al. (2000), Penttinen et al. (2001), LeTerre et al. (2002), Utell et al. (2002), Grahame and Schlesinger (2007), Gong et al. (2005, 2007), Pereira et al. (2007), Araujo et al. (2008), Ayres et al. (2008), Lave and Seskin (1977), Lipfert (1978), Wang et al. (1997), Woodruff et al. (1997), David (2003), Dockery et al. (1989, 1996), Portney et al. (1990), Schwartz (1993), Peters et al. (1997, 1999, 2001), Hoek et al. (1998), Pope et al. (1999, 2004, 2006), Salvi et al. (1999, 2000), Gold et al. (2000), Tan et al. (2000), Peden (2001), Vincent et al. (2001), Brook et al. (2002), Kodavanti et al. (2002), Ghio and Huang (2004), Demeo et al. (2004), Liao et al. (1999, 2004), Schins et al. (2004), Ghio et al. (2000), Park et al. (2005), Schwartz et al. (2005), Rucker et al. (2006), Ulrich et al. (2002), Tarantini et al. (2009), Breton et al. (2011), Peter et al. (2001), Kunzli et al. (2005), Samet and Krewski (2007), Frampton (2001), Stenfors et al. (2004), Schwartz and Dockery (1992), Schwartz and Morris (1995), Schwartz and Zanobetti (2005), Jahn et al. (2011), Rohr and Wyzga (2012), Taner et al. (2013), Hicken et al. (2014), Pascal et al. (2014), Yadav et al. (2014), Rai and Panda (2014), Rai et al. (2014), Raaschou-Nielsen et al. (2013), Beelen et al. (2013), Hamra et al. (2014), Dergham et al. (2015), Kim et al. (2015), Yang et al. (2015), and Rai (2015)

Paticulate matter ($PM_{2.5}$, PM_{10})	PM_{10} exposure is associated with increased ischemic heart disease among the elderly population and with higher risk of myocardial infraction. Increased concentration of dust particulates in the air contribute human health hazards involving acute respiratory disorders such as sinusitis, bronchitis, asthma and allergy, and damage to the defensive functions of alveolar macrophages leading to an increase in respiratory infections.	Beg (1999), Al-Hurban and Al-Ostad (2010), Dockery et al. (1993), and Rai (2015)
Heavy metals in dust particulates	Presence of heavy metals in airborne particulates causes protein denaturation resulting in malfunction or death of cell. It also causes a number of health effects such as cancer, neurotoxicity, immunotoxicity and cardiotoxicity, leading to increased morbidity or mortality in the community.	

After Rai (2015).

Inhalation exposure studies have shown that short-term exposure to diesel exhaust has an acute inflammatory effect on normal human airways resulting in marked neutrophilia, activation of mast cells and neutrophils, and the production of cytokines and chemokines associated with neutrophil accumulation and activation (Salvi et al., 2000; Frampton, 2001; Stenfors et al., 2004; Rai, 2015). Epidemiological studies conducted in different parts of the world have demonstrated an important association between ambient levels of motor vehicle traffic emissions and increased symptoms of asthma and rhinitis (Rai, 2013). Additionally, recent human and animal laboratory-based studies have shown that particulate toxic pollutants, in particular diesel exhaust particles (DEP), can enhance allergic inflammation and induce the development of allergic immune responses (Salvi et al., 2000; Stenfors et al., 2004; Frampton, 2001; Rai, 2015).

2.3 PARTICULATE MATTER (PM) AND DISEASES AFFECTING HUMAN HEALTH

Diesel exhaust-exposed workers have been shown to have an increased risk of lung cancer (Nielsen et al., 1996a,b; Scheepers et al., 2002; Rai, 2015). Methods for the assessment of exposures to diesel exhaust were evaluated by comparing underground workers (drivers of diesel-powered excavators) at an oil shale mine in Estonia with surface workers, and it was observed that underground miners were also occupationally exposed to benzene and polycyclic aromatic hydrocarbons, as indicated by excretion of urinary metabolites of benzene and pyrene, and increased O^6-alkylguanine DNA adducts were detected in the white blood cells of underground workers, suggesting higher exposure to nitroso compounds (Scheepers et al., 2002; Rai, 2015). Diesel exhaust consists of a complex mixture of particulates that contain known genotoxicants, one of which is benzene. Muzyka et al. (1998) indicated significant differences in 5-aminolevulinic acid (ALA) synthesis and heme formation between the exposed workers to PM containing benzene when compared to the nonexposed individuals.

Several air pollutants comprising PM, e.g., benzo[a]pyrene among other carcinogenic polycyclic aromatic hydrocarbons (c-PAHs) and diesel engine exhaust emissions, are classified as class-1 carcinogens, whereas gasoline engine exhaust emissions, carbon black, and a number of PAHs are classified as group 2B for their carcinogenicity by the International Agency for Research on Cancer (IARC) (IARC, 1983, 1987, 1989; Pedersen et al., 2006; Rai, 2015).

Chen et al. (2004) reported that ambient air pollution had acute and chronic effects on mortality, morbidity, hospital admissions, clinical symptoms, lung function changes, etc. in China. Schoket (1999), in his exhaustive study, found that in Silesia, Poland, and Northern Bohemia, Czech Republic, where coal-based industry and domestic heating are the major sources of PAHs, significant differences have been observed in white blood cell DNA adducts and cytogenetic biomarkers between environmentally exposed and rural control populations, and significant seasonal variations of DNA damage have been detected. Further, Schoket (1999) found that in Copenhagen, Athens, Genoa, and Cairo, bus drivers, traffic policemen, and local residents have been involved in biomarker studies and differences have been measured in the levels of DNA damage of urban and rural populations.

Traffic originating from an increased number of vehicles may cause multiple adverse health effects including asthma and allergic diseases, cardiac effects, respiratory symptoms, reduced lung function growth, adverse reproductive outcomes, premature mortality, and lung cancer (White et al., 2005; Samet, 2007). The occurrence of dramatic microsatellite alterations in 3p chromosome multiple critical regions could be a crucial underlying mechanism that exacerbated the lung toxicity in air pollution PM-exposed target L132 cells (Saint-Georges et al., 2009; Rai, 2015).

Traffic policemen are heavily exposed to vehicle exhaust during traffic control and other outdoor activities (Carere et al., 2002), which may lead to increased incidence of sister chromatid exchanges (Anwar, 1994; Chandrasekaran et al., 1996; Zhao et al., 1998; Carere et al., 2002) and micronuclei and chromosomal aberrations (Anwar, 1994; Chandrasekaran et al., 1996; Zhao et al., 1998; Rai, 2015). Carere et al. (2002) while investigating blood cells and DNA of Rome traffic policemen and office workers indicated that exposure to moderate air pollution levels does not result in a detectable increase of genetic damage in blood cells.

Muzyka et al. (1998) demonstrated the data on determination of ALA synthesis and heme formation in lymphocytes from a group of 45 bus-garage workers and analogous data from a group of 25 unexposed subjects, and the outcome indicated significant differences in ALA synthesis and heme formation between the exposed workers when compared to the nonexposed individuals. In addition, concentration of porphyrin associated with DNA was significantly increased (Muzyka et al., 1998). The aforesaid findings of Muzyka et al. (1998) reflect that metabolites of heme synthesis in lymphocytes could be a useful biomonitoring index for the determination of a sensitive subgroup of workers who undergo the higher risk of cancer development.

2.4 SUSPENDED PARTICULATE MATTER (SPM) AND HEALTH RISK

Various studies showed PM exposure was associated with elevated levels of c-reactive protein (CRP), a marker of systemic inflammation that may be an important and independent predictor of cardiovascular disease. For example, a recent study reported associations between CRP and interleukin (IL)-6 with PM in subjects with coronary artery disease (Delfino et al., 2008; Rai, 2015), inflammatory lung injury, bone marrow and blood cell responses, enhanced human alveolar macrophage production of proinflammatory cytokines, elevated blood plasma viscosity (Ghio and Huang, 2004; Rai, 2015), endothelial dysfunction and brachial artery vasoconstriction, and triggering of myocardial infarction. Polichetti et al. (2009) extensively reviewed the impact of PM on the cardiovascular system. Particulate matter is also linked with psychosocial stress and high blood pressure (Hicken et al., 2014; Rai, 2015).

Suspended particulate matter (SPM) is of the greatest concern as it contributes 50% to total air pollution and causes respiratory disorders in human beings on prolonged exposure (Freer-Smith et al., 2004) as it include all airborne particles in the size range of 0.5–100 μm. Sirajuddin and Ravichandran (2010) also studied SPM-related respiratory disorders such as nose block, sneezing, cough, and hyperacidity in Tiruchirappalli, India. Bhattacharjee et al. (2010) noted that $PM < 4.6\,\mu m$ and $PM < 1.1\,\mu m$ are hazardous to human health due to their capacity to be inhaled into the bronchial region and be deposited in the alveolar region. Air quality monitoring at Kolkata, India (for SPM, RPM, NO_x, SO_2, CO, and Pb levels) indicated that they are currently at levels dangerous to human health (Ghose et al., 2005; Rai, 2015).

It is well documented in the literature that particulate pollution causes adverse health impact particularly in the size range of less than 10 μm (Curtis et al., 2006; Lipmann, 2007; Zeger et al., 2008; Mitchell et al., 2010; Rai, 2015). PM pollutants are associated with adverse effects on the respiratory system (Schwartz, 1996; Pope et al., 2002; Knutsen et al., 2004; Knox, 2006; Maher et al., 2008; Hansard et al., 2011; Rai, 2015). If these particulates of size smaller than 10 μm cause inflammation, diminished pulmonary function can be unavoidable (Knutsen et al., 2004; Seaton et al., 1995; Maher et al., 2008; Rai, 2015). Further, PM with aerodynamic diameter smaller than

2.5 μm (PM$_{2.5}$) have even more deleterious health impacts because when inhaled they penetrate deeper than PM$_{10}$ and can reach the lungs' alveoli (Rizzio et al., 1999; Harrison and Yin, 2000; Wichmann and Peters, 2000; Saragnese et al., 2011; Rai, 2015). Links with lung cancer (Pope et al., 2002) and increased cardiovascular mortality rates (Schwartz, 1996) have also been established. Lung diseases due to PM may be attributed to the presence of inflammatory cells in the airways including neutrophils, eosinophils, and monocytes, and increased numbers of alveolar macrophages (Becker et al., 2002; Rai, 2015).

Children are particularly sensitive to air pollution as their lungs as well as immune systems are not completely developed (Bateson and Schwartz, 2007). Furthermore, children are considered more vulnerable to the adverse effects of air pollution than adults due to physiological differences related to their body size, growth, development, and immaturity of organs and body functions (Pedersen et al., 2006). The air intake of a resting infant less than one year old is twice that of an adult (Pedersen et al., 2006). Children also exhibit a higher intake of food and water per kilogram body weight (Pedersen et al., 2006). Bener et al. (2007) determined the impact of asthma and air pollution on school attendance of primary school children aged 6–12 years in Qatar and found that air pollution has an impact on asthma, which results in significant school absenteeism. Wilhelm et al. (2005) assessed the impact of Pb and Cs in child–mother pairs in North Rhine Westphalia, Germany, and through regression analysis showed that Pb levels in ambient air were associated with Pb in the blood of children and mothers.

2.5 HUMAN HEALTH CASE STUDIES AND DATA

Urinary 1-hydroxypyrene (1-OHP), a major metabolite of pyrene, may act as a biomarker of exposure to PAHs (Jonganeelen et al., 1990; Bouchard and Viau, 1999; Cavanagh et al., 2007; Rai, 2015). Cavanagh et al. (2007) found 1-OHP in elevated concentrations in school children immediately after heavy exposure of particulate pollution, particularly PAH.

Global records show that PM less than 2.5 μm causes 3% of mortality from cardiopulmonary disease; 5% of mortality from cancer of the trachea, bronchus, and lung; and 10% of mortality from acute respiratory infections in children under age five (Cohen et al., 2005; Maher, 2009; Rai, 2015). It is well established in the literature that air pollution with PM in children results in detectable effects indicated by a number of biomarkers of exposure and early effects (Pedersen et al., 2006; Rai, 2015). The aforesaid hypothesis was tested through a family pilot study that was conducted in the Czech Republic through fluorescence in situ hybridization; it was concluded that micronuclei are a valuable and sensitive biomarker for early biological effects in children and adults living in two different areas characterized by significant exposure differences in c-PAH concentrations during the winter (Pedersen et al., 2006; Rai, 2015).

PM pollution is intimately linked with cardiovascular morbidity and mortality (Brook et al., 2004; Adar and Kaufman, 2007). Several research studies demonstrated that approximately 50% of the adult mortality from air pollution, or 22,000 deaths per year, were due to traffic sources in Austria, France, and Switzerland (Kunzli et al., 2000; Adar and Kaufman, 2007; Rai, 2015). Adar and Kaufman (2007) reviewed the epidemiological evidence regarding the impact of traffic-related pollution on cardiovascular diseases and ongoing epidemiological studies are underway to identify the cardiovascular health impacts of traffic.

Further, several studies have shown an association between exposure to ambient PM and hospital admissions for cardiac and respiratory causes (Analitis et al., 2006; Atkinson et al., 2001; Le Tertre et al., 2002; Migliaretti et al., 2005; Sun et al., 2006; Ciencewicki and Jaspers, 2007). Moreover, it has been shown that hospital visits for asthma associated with exposure to PM are more prevalent in children and the elderly (Migliaretti et al., 2005; Sun et al., 2006; Ciencewicki and Jaspers, 2007) and that exposure to PM is associated with decreased lung function in children (Moshammer et al., 2006; Ciencewicki and Jaspers, 2007; Rai, 2015).

2.6 HUMAN HEALTH RISK DUE TO PAH AND VOCs

In general, PM comprises polycyclic aromatic hydrocarbons and volatile organic compounds (VOCs), which may have deleterious impacts on human health. Air pollutants, particularly VOCs, have been reported from waste treatment and disposal facilities and may be of concern to public health (Hamoda, 2006). Associations between urban pollutants and respiratory and cardiovascular problems, and still a greater incidence of certain cancer types, have already been mentioned in the literature (Lester and Seskin, 1970; Saldiva et al., 2002; Lin et al., 2003; Rai, 2015).

Ciencewicki and Jaspers (2007) reviewed the association between and effects of air pollutants and respiratory viral infections, as well as potential mechanisms associated with these effects. PM, especially traffic-related airborne particles, contains a large number of genotoxic/mutagenic chemical substances, which can cause DNA damage and promote malignant neo-plasms (Valavanidis et al., 2008; Rai, 2015). The genotoxicity of PM was extensively studied with the *Salmonella typhimurium* assay (Ames test) by various research groups and reviewed by Claxton et al. (2004) and sum-marized elsewhere (Valavanidis et al., 2008). Combustion emissions account for over half of the fine particle ($PM_{2.5}$) air pollution and most of the pri-mary particulate organic matter (Lewtas, 2007). Lewtas (2007) in his exten-sive review on the impacts of air pollution on human beings mentioned that both short- and long-term exposures to combustion emissions and ambient fine particulate air pollution have been associated with measures of genetic damage. Moreover, long-term epidemiological studies have reported an increased risk of all causes of mortality, cardiopulmonary mor-tality, and lung cancer mortality associated with increasing exposures to air pollution. Adverse reproductive effects (e.g., risk for low birth weight) have also recently been reported in Eastern Europe and North America (Lewtas, 2007). Study of a nonsmoking female population in Silesia with a limited age range and a homogeneous occupation indicated that environmental exposure to air pollution may be responsible for genetic damage (Michalska et al., 1999; Rai, 2015).

2.7 MECHANISM ASSOCIATED WITH PAH AND VOCs

Constituents of PM pollutants emanating from vehicular emissions have been demonstrated to cause genotoxicological impact on plants as well as humans. For instance, although PAHs are relatively chemically inert compounds, through metabolic activation to electrophilic derivatives (e.g., diolepoxides, quinones, conjugated hydroxyalkyl derivatives) these are capable of covalent interaction with nucleophilic centers of DNA (Schoket, 1999; Rai, 2015). These adducts of PAH to DNA cause base pair substitutions, frameshift mutations, deletions, S-phase arrest, strand break-age, and a variety of chromosomal alterations (Schoket, 1999). Two main cell types are likely to interact with inhaled particles i.e., alveolar mac-rophages and airway epithelial cells in response to PM pollution (Becker et al., 2005; Rai, 2015). Dagher et al. (2006) demonstrated that in vitro short-term exposure to $PM_{2.5}$ induced oxidative stress and inflammation in human lung epithelial cells (L132) and emphasized the need for such research to reveal the mechanism of adverse health impact imposed by PM. PM induces the activity of NF-kB manifold, which is a transcription factor that can induce gene transcription in a variety of proinflammatory cytokines, enzymes that generate mediators of inflammation and immune receptors (Yang and Omaye, 2009). Research studies have investigated the biological effects of $PM_{2.5}$ on human lung epithelial cell line A549 (Calcabrini et al., 2004). Billet et al. (2008) assessed the genotoxic potential of PAH containing PM on human lung epithelial A549 cells. Fine PM is also reported to induce sister chromatid exchange in human tracheal epithelial cells (Hornberg et al., 1998; Rai, 2015).

A statistically significant increase was established in the frequency of chromosomal aberrations in peripheral blood lymphocytes from the exposed population toward heavy metals and dioxins/furans, hence, chromosomal aberrations in human peripheral blood lymphocytes generally may be used as a biomarker (Huttner et al., 1999). Hellman et al. (1999) demonstrated that increased levels of radon in indoor air ($>200 Bq/m^3$) were found to be associated with an increased level of DNA damage in peripheral lymphocytes.

Reports indicated that constituents of inhaled PM may trigger a proinflammatory response in nervous tissue that could contribute to the pathophysiology of neurodegenerative diseases (Campbell et al., 2006). PM affects sensory and neural pathways through activation of capsaicin-sensitive vanilloid irritant receptors (Veronesi and Oortgiesen, 2001; Rai, 2015).

Claxton and Woodall (2007) for the first time extensively reviewed the mutagenicity and carcinogenicity of air pollutants. Coronas et al. (2009) investigated genotoxic effects on people exposed to an oil refinery in southern Brazil and the mutagenic activity of airborne PM. Individuals who were environmentally exposed to heavy metals (mercury and lead), organic (styrene, formaldehyde, phenol, and benzo[a]pyrene) and inorganic (sulfur and nitrogen oxides, hydrogen, and ammonium fluorides) volatile substances, may have high rates of chromosomal aberrations and sister-chromatid exchanges (Lazutka et al., 1999).

In an integrated way, PAHs and VOCs lead to the formation of bulky DNA adducts (Moller et al., 2008). Figure 2.1 represents the mechanism through which PM leads to formation of DNA adducts. The genotoxic effect of pollutants on the ecosystem, including the buildup of resistant species, is also of considerable concern (Ma et al., 1994; Grant, 1998). The potential genotoxic effects on human health by such vapor phase chemicals include malignant cell formation, the accumulation of heritable abnormal genes within the population, heart disease, aging, and cataracts (Grant, 1998). The effects of toxic compounds, and the subsequent genotoxic effects on plants, are of particular importance as plants comprise a large portion of

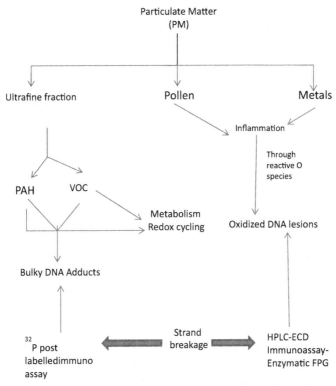

Figure 2.1 Interaction of PM and its constituents with DNA leading to DNA adduct formation. *After Rai (2015).*

our biosphere and constitute a vital link in the food chain (Grant, 1998; Rajput and Agrawal, 2005). Estimating air genotoxicity is therefore crucial to evaluating risk to the environment and public health. PAHs that are deleterious components of particulates result in risk of breast cancer (Gammon and Santella, 2008). Further, PAHs cause DNA damage to T- and B-lymphocytes and granulocytes (specifically single-strand DNA breakage) in individuals exposed to exhaust fumes (Sul et al., 2003).

As stated earlier, PMs also consist of heavy metals, carbon core, and pollen, which further interact with gaseous pollutants resulting in the formation of DNA lesions. PM (specifically PM_{10} and $PM_{2.5}$) contains several deleterious metals. The problem of metals is very much prevalent in the atmosphere, particularly in the Asian countries that were extensively reviewed by Fang et al. (2005). Pb may have several genetic impacts in human beings, e.g., fidelity of DNA synthesis, mutations, chromosomal aberrations, cancers, and birth defects (Johnson, 1998). It reacts or complexes with many biomolecules and adversely affects the reproductive, nervous, gastrointestinal, immune, renal, cardiovascular, skeletal, muscular, and hematopoietic systems, as well as developmental processes (Johnson, 1998).

Research has shown that the size of the airborne particles and their surface area determine the potential to elicit inflammatory injury, oxidative damage, and other biological effects (Valavanidis et al., 2008; Rai, 2015).

There has been considerable concern about the pulmonary effects of particulates less than 2.5 µm ($PM_{2.5}$) or 10 µm (PM_{10}), as they can reach the alveoli and translocate to the circulation, whereas particles of larger size deposit mainly in the upper airways and can be cleared by the mucociliary system (Oberdorster et al., 2005; Moller et al., 2008). Recently, many studies have highlighted the role of ambient airborne PM as an important environmental pollutant for many different cardiopulmonary diseases and lung cancer (Valavanidis et al., 2008). Further, researchers have increasingly realized that generation of reactive oxygen species (ROS) and oxidative stress is an important toxicological mechanism of particle-induced lung cancer (Knaapen et al., 2004; Risom et al., 2005). The fraction of PM contains a number of constituents that may increase the generation of ROS by a variety of reactions such as transition metal catalysis, metabolism, redoxcycling of quinones, and inflammation. PM, thus, can generate oxidative damage to DNA, including guanine oxidation, which is mutagenic (Kasai, 1997; Moller et al., 2008). The oxidative stress mediated by PM and resulting DNA damage may originate from generation of ROS from the surface of particles, soluble compounds such as transition metals or organic compounds, altered function of mitochondria or NADPH-oxidase, and activation of inflammatory cells capable of generating ROS and reactive nitrogen species (Risom et al., 2005). Production of ROS and the secretion of inflammatory cytokines could interact by inducing cell death by apoptosis (Shukla et al., 2000; Haddad, 2004; Hetland et al., 2004; Dagher et al., 2006) (Figure 2.2). Interaction between oxidative stress by-products and certain genes within our population may modulate the expression of specific chronic diseases (Yang and Omaye, 2009; Rai, 2015).

Epidemiological studies have observed positive associations between levels of PM and the incidence of mortality, including those involving cardiovascular and respiratory conditions (Samet et al., 2000a; Katsouyanni et al., 2001; Pope et al., 2002; Dominici et al., 2003; Ciencewicki and Jaspers, 2007; Rai, 2015). Likewise, numerous epidemiological studies have highlighted the health implications of fine particles with aerodynamic diameter smaller than 10 µm (Kunzli et al., 2000; Katsouyanni et al., 2001; Pandey et al., 2005, 2006; Pope et al., 2002;

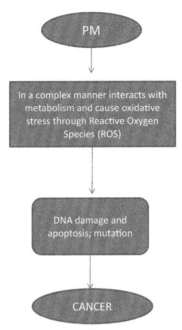

Figure 2.2 Impact of PM on cellular machinery leading to apoptosis. *After Rai (2015).*

Peng et al., 2005; Rai, 2015). Epidemiological findings suggest that short-term PM exposure can trigger acute or terminal health events whereas long-term PM exposure can promote life-shortening chronic illness. PM was also associated with decreases in DNA methylation in *NOS2A*, a gene directly responsible for production of nitric oxide, an important player in both respiratory and cardiovascular diseases (Tarantini et al., 2009; Breton et al., 2011). A study in China reported a significant adverse relationship between birth weight and maternal exposure to total SPM (Wang et al., 1997) suggesting that both pre and postnatal exposure to ambient PM affects infant birth weight and mortality. Other studies have also suggested a significant link between PM exposure and excess infant mortality (Lave and Seskin, 1977; Lipfert, 1978; Woodruff et al., 1997). Woodruff et al. (1997) also studied the link between PM_{10} exposures to an increased incidence of sudden infant death syndrome. Additional evidence suggested that PM exposure over time can alter lung function, lung tissue and structure, airway responsiveness and respiratory defense mechanisms, and can increase susceptibility to respiratory infection and damage respiratory cells (EPA, 1996, 1997). The association of ambient PM with mortality and cardiovascular outcome has been well established in studies of short-term exposure (Levy et al., 2000; Samet et al., 2000a,b; Zanobetti et al., 2000). $PM_{2.5}$

exposure increases ischemic cardiovascular events and promotes atherosclerosis (Samet and Krewski, 2007; Pope et al., 2006).

Christophe et al. (1997) have reported some metals in airborne dust particulates as having toxic effects on human health. Heavy metals associated with respirable dust particles below $10\,\mu m$ (PM_{10}) in urban air can penetrate deep into the lungs, causing many health problems. Heavy metals are toxic even at low concentrations in air. After they are inhaled, they form complexes or legends with vital protein molecules, denaturing them and resulting in malfunction or cell death (USEPA, 1996). Heavy metals are now correlated with a number of health effects such as cancers, neurotoxicity, immunotoxicity and cardiotoxicity, leading to increased morbidity or mortality in humans (Dockery et al., 1993; Rai, 2015).

Beg (1999) performed a study on the toxic response of urban dust leading to the dispersion of PM in the atmosphere at a shocking high concentration. Increased concentration of this inhalable PM in the air due to dust storms contributes human health hazards including acute respiratory disorders such as sinusitis, bronchitis, asthma and allergy, and damage to the defensive functions of alveolar macrophages leading to increase respiratory infections. Research has also found that exposure to ambient PM concentrations results in increased blood level of endothelins, which can affect vascular tone and endothelial function (Vincent et al., 2001; Brook et al., 2002).

In the early and mid-1990s, studies by Schwartz and his colleagues (Schwartz and Dockery, 1992; Schwartz and Morris, 1995) found increased ischemic heart disease among the elderly population associated with PM_{10} exposure. Schwartz and Zanobetti (2005) also observed that $PM_{2.5}$ and PM_{10} were strongly associated with higher risk of myocardial inflammation. Al-Hurban and Al-Ostad (2010) conducted a study on the textural and mineralogical characteristics of the dust fallout and their effect on human health in Kuwait city. It was found that the dust was mostly comprised of a high concentration of calcite and quartz with grain size ranging $1-25\,\mu m$. Inhalation of calcite particles may cause coughing, sneezing, and nasal irritation, and in cases of chronic exposure to excessive oral doses of calcite may produce alkalosis and hypercalcemia. Quartz can have potentially serious respiratory effects following long-term exposure; in particular, it is a known carcinogen.

Experimental data show that redox active PM components (that are especially enriched in ultrafine $PM < 0.1\,\mu m$) lead to the production of ROS in various cells in the lungs, blood, and vascular tissues. This is followed by oxidative stress, which can then lead to increased airway and systemic inflammation, and adverse cardiovascular responses when antioxidant defenses are overwhelmed (Ayres et al., 2008; Utell et al., 2002; Rai, 2015). Oxidative stress is a biochemical imbalance in which production of ROS exceeds the natural antioxidant capacity. This imbalance can occur in the body following exposure to pro-oxidant air pollutants. Oxidative stress may play a central role in the respiratory and cardiovascular effects of air pollution through its immune modulating effects and its ability to initiate the inflammatory process and thrombogenic activity.

REFERENCES

Adar, S.D., Kaufman, J.D., 2007. Cardiovascular disease and air pollutants: evaluating and improving epidemiological data implicating traffic exposure. Inhalation Toxicology 19 (1), 135–149.

Al-Hurban, A., Al-Ostad, A., 2010. Textural characteristics of dust fallout and potential effect on public health in Kuwait city and suburbs. Environmental Earth Sciences 60, 169–181.

Analitis, A., Katsouyanni, K., Dimakopoulou, K., Samoli, E., Nikoloulopoulos, A.K., Petasakis, Y., Touloumi, G., Schwartz, J., Anderson, H.R., Cambra, K., Forastiere, F., Zmirou, D., Vonk, J.M., Clancy, L., Kriz, B., Bobvos, J., Pekkanen, J., 2006. Short-term effects of ambient particles on cardiovascular and respiratory mortality. Epidemiology 17, 230–233.

Anwar, W.A., 1994. Monitoring of human populations at risk by different cytogenetic end points. Environmental Health Perspectives 102 (4), 131–134.

Araujo, J.A., Barajas, B., Kleinman, M., et al., 2008. Ambient particulate pollutants in the ultrafine range promote early atherosclerosis and systemic oxidative stress. Circulation Research 102, 589–596.

Atkinson, R.W., Anderson, H.R., Sunyer, J., Ayres, J., Baccini, M., Vonk, J.M., Boumghar, A., Forastiere, F., Forsberg, B., Touloumi, G., Schwartz, J., Katsouyanni, K., 2001. Acute effects of particulate air pollution on respiratory admissions: results from APHEA 2 project. Air pollution and health: a European approach. American Journal of Respiratory and Critical Care Medicine 164, 1860–1866.

Ayres, J.G., Borm, P., Cassee, F.R., et al., 2008. Evaluating the toxicity of airborne particulate matter and nanoparticles by measuring oxidative stress potential—a workshop report and consensus statement. Inhalation Toxicology 20, 75–99.

Bateson, T.F., Schwartz, J., 2007. Children's response to air pollutants. Journal of Toxicology and Environmental Health Part A 71 (3), 238–243.

Becker, S., Mundandhara, S., Devlin, R.B., Madden, M., 2005. Regulation of cytokine production in human alveolar macrophages and airway epithelial cells in response to ambient air pollution particles: further mechanistic studies. Toxicology and Applied Pharmacology 207, S269–S275.

Becker, S., Soukup, J.M., Gallagher, J.E., 2002. Differential particulate air pollution induced oxidant stress in human granulocytes, monocytes and alveolar macrophages. Toxicology In Vitro 16, 209–218.

Beelen, R., Raaschou-Nielsen, O., Stafoggia, M., et al., 2013. Effects of long-term exposure to air pollution on natural-cause mortality: an analysis of 22 European cohorts within the multicentre ESCAPE project. The Lancet 383, 785–795.

Beg, M.U., 1999. An overview of toxic response of urban dust. In: Proceedings of the International Conference Pollution. Environmental Pollution. 99, 347–360.

Bener, A., Kamal, M., Shanks, J., 2007. Impact of asthma and air pollution on school attendance of primary school children: are they at increased risk of school absenteeism? Journal of Asthma 44 (4), 249–252.

Bhattacharjee, A., Mandal, R., Chini, K.T., 2010. A preliminary study on the nature of particulate matters in vehicle fuel wastes. Environmental Monitoring and Assessment. http://dx.doi.org/10.1007/S10661-010-1598-X.

Billet, et al., 2008. Genotoxic potential of polycyclic aromatic hydrocarbons-coated onto airborne Particulate Matter (PM2.5) in human lung epithelial A549 cells. Cancer Letters 270, 144–155.

Bouchard, M., Viau, C., 1999. Urinary 1-hydroxypyrene as a biomarker to polycyclic aromatic hydrocarbons: biological monitoring strategies and methodology for determining biological exposure indices for various work environments. Biomarkers 4, 139–187.

Breton, C.V., Byun, H.M., Wang, X., Salam, M.T., Siegmund, K., Gilliland, F.D., 2011. PM2.5 is associated with DNA methylation in inducible nitric oxide synthase. [abstract]. In: American Thoracic Society International Conference.

Brook, R.,D., Brook, J.R., Urch, B., Vincent, R., Rajagopalan, S., Silverman, F., 2002. Inhalation of fine particulate air pollution and ozone causes acute arterial vasoconstriction in healthy adults. Circulation 105, 1534–1536.

Brook, R.D., Franklin, B., Cascio, W., Hong, Y.L., Howard, G., Lipsett, M., et al., 2004. Air pollution and cardiovascular diseasea statement for healthcare professionals from the expert panel on population and prevention science of the American Heart Association. Circulation 109 (21), 2655–2671.

Calcabrini, et al., 2004. Fine environmental particulate engenders alterations in human lung epithelial A549 cells. Environmental Research 95, 82–91.

Campbell, et al., 2006. Particulate matter in polluted air may increase biomarkers of inflammation in mouse brain. Neurotoxicology 26, 133–140.

Carere, et al., 2002. Biomonitoring of exposure to urban air pollutants: analysis of sister chromatid exchanges and DNA lesions in peripheral lymphocytes of traffic policemen. Mutation Research 518, 215–224.

Cavanagh, J.E., Brown, L., Trought, K., Kingham, S., Epton, M.J., 2007. Elevated concentrations of 1-hydroxypyrene in school children during winter in Christchurch, New Zealand. Science of the Total Environment 374, 51–59.

Chandrasekaran, R., Samy, P.L., Murthy, P.B., 1996. Increased sister chromatid exchange (SCE) frequencies in lymphocytes from traffic policemen exposed to automobile exhaust pollution. Human Experimental Toxicology 15, 301–304.

Chen, B., Hong, C., Kan, H., 2004. Exposures and health outcomes from outdoor air pollutants in China. Toxicology 198, 291–300.

Christophe, M., Blandine, J., Emmanuel, N., 1997. Measurement of trace metals in wet, dry and total atmospheric fluxes over the Ligurian Sea. Atmospheric Environment 6, 889–896.

Ciencewicki, J., Jaspers, I., 2007. Air pollution and respiratory viral infection. Inhalation Toxicology 19 (14), 1135–1146.

Claxton, L.D., Matthews, P.P., Warren, S.S., 2004. The genotoxicity of ambient outdoor air: a review: Salmonella mutagenicity. Mutation Research 567, 347–399.

Claxton, L.D., Woodall, G.M.J., 2007. A review of the mutagenicity and rodent carcinogenicity of ambient air. Mutation Research/Reviews in Mutation Research 636 (1–3), 36–94.

Cohen, A.J., Anderson, H.R., Ostro, B., Pandey, K.D., Krzyzanowski, M., Kuenzli, N., 2005. The global burden of disease due to outdoor air pollution. Journal of Toxicology and Environmental Health, Part A 68 (13–14), 1301–1307.

Coronas, et al., 2009. Genetic biomonitoring of an urban population exposed to mutagenic airborne pollutants. Environment International 35 (7), 1023–1029.

Curtis, L., Rea, W., Smith-Willis, P., Fenyves, E., Pan, Y., 2006. Adverse health effects of outdoor pollutants. Environment International 32, 815–830.

Dagher, Z., et al., 2006. Activation of different pathways of apoptosis by air pollution Particulate Matter ($PM_{2.5}$) in human epithelial lung cells (L132) in culture. Toxicology 225, 12–24.

David, D.-S., Lidia, P., Riccardo, P., 2003. Current allergy and asthma reports. Current Science 3, 146–152.

Delfino, R.J., Staimer, N., Tjoa, T., Polidori, A., Arhami, M., Gillen, D.L., Kleinman, M.T., Vaziri, N.D., Longhurst, J., Zaldivar, F., Sioutas, C., 2008. Circulating biomarkers of inflammation, antioxidant activity, and platelet activation are associated with primary combustion aerosolsin subjects with coronary artery disease. Environmental Health Perspectives 116 (7), 898–906.

DeMeo, D.L., Zanobetti, A., Litonjua, A.A., Coull, B.A., Schwartz, J., Gold, D.R., 2004. Ambient air pollution and oxygen saturation. American Journal Respiratoryand Critical Care Medicine 170, 383–387.

Dergham, M., et al., 2015. Temporal–spatial variations of the physicochemical characteristics of air pollution Particulate Matter ($PM_{2.5-0.3}$) and toxicological effects in human bronchial epithelial cells (BEAS-2B). Environmental Research 137, 256–267.

Dockery, D.W., Cunningham, J., Damokosh, A.I., Neas, L.M., Spengler, J.D., Koutrakis, P., Ware, J.H., Raizenne, M., Speizer, F.E., 1996. Health effects of acid aerosols on North American children: respiratory symptoms. Environmental Health Perspectives 104, 500–505.

Dockery, D.W., Pope, C.A., Xu, X., Spengler, J.D., Ware, J.H., Fay, M.E., Ferris, B.G., Speizer, F.E., 1993. An association between air pollution and mortality in six US cities. New England Journal of Medicine 329, 1753–1759.

Dockery, D.W., Speizer, F.E., Stram, D.O., Ware, J.H., Spengler, J.D., Ferris, B.G., 1989. Effects of inhalable particles on respiratory health of children. The American Review of Respiratory Disease 139, 587–594.

Dominici, F., McDermott, A., Zeger, S.L., Samet, J.M., 2003. Airborne particulate matter and mortality: timescale effects in four US cities. American Journal of Epidemiology 157, 1055–1065.

Donaldson, K., Stone, V., Gilmour, P.S., Brown, D.M., MacNee, W., 2000. Ultrafine particles: mechanisms of lung injury. Philosophy Transactions of the Royal Society of London A358, 2741–2749.

Driscoll, K.E., Carter, J.M., Hassenbein, D.G., Howard, B., 1997. Cytokines and particle-induced inflammatory cell recruitment. Environmental Health Perspectives 105, 1159–1164.

Environmental Protection Agency (U.S.), 1996. Air Quality Criteria for Particulate Matter. Office of Research and Development, Office of Health and Environmental Assessment, Research Triangle Park, NC. EPA report no. EPA/600/P-95/001aF, USA.

Environmental Protection Agency (U.S.), 1997. Regulatory Impact Analysis for the Particulate Matter and Ozone. National Ambient Air Quality Standards and Proposed Regional Haze Rule. Innovative Strategies and Economics Group, Office of Air Quality Planning and Standards, Research Triangle Park, NC.

Fang, G.C., Wu, Y.S., Huang, S.H., Rau, J.Y., 2005. Review of atmospheric metallic elements in Asia during 2000–2004. Atmospheric Environment 39 (17), 3003–3013.

Frampton, M.W., 2001. Systematic and cardiovascular effects of airway injury and inflammation: ultrafine particle exposure in humans. Environmental Health Perspectives 109, 529–532.

Freer-Smith, P.H., El-Khatib, A.A., Taylor, G., 2004. Capture of particulate pollution by trees: a comparison of species typical of semi-arid areas (Ficus nitida and Eucalyptus globulus) with European and North-American species. Water, Air, and Soil Pollution 155, 173–187.

Gammon, M.D., Santella, R.M., 2008. PAH, genetic susceptibility and breast cancer risk: an update from the Long Island Breast Cancer Study Project. European Journal of Cancer 44 (5), 636–640.

Ghio, A.J., Huang, Y.T., 2004. Exposure to concentrated ambient particles (CAPs): a review. Inhalation Toxicology 16, 53–59.

Ghio, A.J., Kim, C., Devlin, R.B., 2000. Concentrated ambient air particles induce mild pulmonary inflammation in healthy human volunteers. American Journal of Respiratoryand Critical Care Medicine 162, 981–988.

Ghose, M.K., Paul, R., Banerjee, S.K., 2005. Assessment of the impact on human health of exposure to urban air pollutants: an Indian case study. International Journal of Environmental Studies 62 (2), 201–214.

Gold, D.R., Litonjua, A., Schwartz, J., Lovett, E., Larson, A., Nearing, B., Allen, G., Verrier, M., Cherry, R., Verrier, R., 2000. Ambient pollution and heart rate variability. Circulation 101, 1267–1273.

Gong Jr., H., Linn, W.S., Clark, K.W., Anderson, K.R., Geller, M.D., Sioutas, C., 2005. Respiratory responses to exposures with fine particulates and nitrogen dioxide in the elderly with and without COPD. Inhalation Toxicology 17, 123–132.

Gong, K.W., Zhao, W., Li, N., et al., 2007. Air-pollutant chemicals and oxidized lipids exhibit genome-wide synergistic effects on endothelial cells. Genome Biology 8, R149.

Grahame, T.J., Schlesinger, R.B., 2007. Health effects of airborne particulate matter: do we know enough to consider regulating specific particle types or sources? Inhalation Toxicology 19, 457–481.

Grant, W.F., 1998. Higher plant assays for the detection of genotoxicity in air polluted environments. Ecosystem Health 4, 210–229.

Haddad, J.J., 2004. Redox and oxidant-mediated regulation of apoptosis signaling pathways: immuno-pharmaco-redox conception of oxidative siege versus cell death commitment. International Immunopharmacology 4, 475–493.

Hamoda, M.F., 2006. Air pollutants emissions from waste treatment and disposal facilities. Journal of Environmental Science and Health, Part A 41 (1), 77–85.

Hamra, G.B., Guha, N., Cohen, A., Laden, F., Raaschou-Nielsen, O., Samet, J.M., Vineis, P., Forastiere, F., Saldiva, P., Yorifuji, T., Loomis, D., 2014. Outdoor particulate matter exposure and lung cancer: a systematic review and meta-analysis. Environmental Health Perspectives 212 (9), 906–911.

Hansard, R., Maher, B.A., Kinnersley, R., 2011. Biomagnetic monitoring of industry-derived particulate pollution. Environmental Pollution 159, 1673–1681.

Harrison, R.M., Yin, J., 2000. Particulate matter in the atmosphere: which particle properties are important for its effects on health? The Science of the Total Environment 249, 85–101.

Hellman, B., Friis, L., Vaghef, H., Edling, C., 1999. Alkaline single cell gel electrophoresis and human biomonitoring for genotoxicity: a study on subjects with residential exposure to radon. Mutation Research 442, 121–132.

Hetland, R.B., Cassee, F.R., Refsnes, M., Schwarze, P.E., Lag, M., Boere, A.J.F., Dybing, E., 2004. Release of inflammatory cytokines, cell toxicity and apoptosis in epithelial lung cells after exposure to ambient air particles of different size fractions. Toxicology In Vitro 18, 203–212.

Hicken, M.T., et al., 2014. Fine particulate matter air pollution and blood pressure: the modifying role of psychosocial stress. Environmental Research 133, 195–203.

Hoek, G., Dockery, D.W., Pope III, C.A., Neas, L., Roemer, W., Brunekreef, B., 1998. Association between PM10 and decrements in peak expiratory flow rates in children: reanalysis of data from 5 panel studies. European Respiratory Journal 11, 1307–1311.

Hornberg, C., Maciuleviciute, L., Seemayer, N.H., Kainka, E., 1998. Induction of sister chromatid exchanges (SCE) in human tracheal epithelial cells by the fractions PM-10 and PM-2.5 of airborne particulates. Toxicology Letters 96 (97), 215–220.

Huttner, E., Gotze, A., Nikolova, T., 1999. Chromosomal aberrations in humans as genetic endpoints to assess the impact of pollution. Mutation Research 445, 251–257.

International Agency for Research on Cancer, 1983. Evaluation of Carcinogenic Risk of Chemicals to Humans. Polycyclic Aromatic Compounds, Part 1. Chemical, Environmental and Experimental Data. Monograph No. 32. IARC, Lyon, France.

International Agency for Research on Cancer, 1987. Evaluation of Carcinogenic Risk to Humans. Overall Evaluations of Carcinogenicity: an Updating of IARC. Monographs Volumes 1–42, Suppl. 7. IARC, Lyon, France.

International Agency for Research on Cancer, 1989. Evaluation of Carcinogenic Risk of Chemicals to Humans, Diesel and Gasoline Engine Exhausts and Some Nitroarenes. Monograph No. 46. IARC, Lyon, France.

Jahn, H.J., et al., 2011. Particulate matter pollution in the megacities of the Pearl River Delta, China–a systematic literature review and health risk assessment. International Journal of Hygiene and Environmental Health 214 (4), 281–295.

Johnson, F.M., 1998. The genetic effects of environmental lead. Mutation Research/Reviews in Mutation Research 410 (2), 123–140.

Jonganeelen, F.J., van Leeuwen, F.E., Oosterink, S., Anzion, R.B.M., van der Loop, F., Bos, R.P., 1990. Ambient and biological monitoring of cokeoven workers: determinants of the internal dose of polycyclic aromatic hydrocarbons. British Journal of Industrial Medicine 4, 454–461.

Kasai, H., 1997. Analysis of a form of oxidative DNA damage, 8-hydroxy-20-deoxyguanosine, as a marker of cellular oxidative stress during carcinogenesis. Mutation Research 387, 147–163.

Katsouyanni, K., Touloumi, G., Samoli, E., et al., 2001. Confounding and effect modification in the short-term effects of ambient particles on total mortality: results from 29 European cities within the APHEA2 project. Epidemiology 12, 521–531.

Kim, K., Kabir, E., Kabir, S., 2015. A review on the human health impact of airborne particulate matter. Environment International 74, 136–143.

Knaapen, A.M., Borm, P.J., Albrecht, C., Schins, R.P., 2004. Inhaled particles and lung cancer. Part A: mechanisms. International Journal of Cancer 109, 799–809.

Knox, E.G., 2006. Roads, railways and childhood cancers. Journal of Epidemiology and Community Health 60, 136–141.

Knutsen, S., Shavlik, D., Chen, L.H., Beeson, W.L., Ghamsary, M., Petersen, F., 2004. The association between ambient particulate air pollution levels and risk of cardiopulmonary and all-cause mortality during 22 years follow-up of a non-smoking cohort. Results from the AHSMOG study. Epidemiology 15, S45.

Kodavanti, U.P., Schladweiler, M.C., Ledbetter, A.D., Hauser, R., Christiani, D.C., McGee, J., Richards, J.R., Costa, D.L., 2002. Temporal association between pulmonary and systemic effects of particulate matter in healthy and cardiovascular compromised rats. Journal of Toxicology and Environmental Health A 65, 1545–1569.

Kunzli, N., Jerret, M., Mack, W.J., Beckerman, B., LaBree, L., Gilliland, F., Thomas, D., Peters, J., Hodis, H.N., 2005. Ambient air pollution and atherosclerosis in Los Angeles. Environmental Health Perspectives 113, 201–206.

Kunzli, N., Kaiser, R., Medina, S., Studnicka, M., Chanel, O., Filliger, P., et al., 2000. Public-health impact of outdoor and traffic-related air pollution: a European assessment. The Lancet 356 (9232), 795–801.

Lave, L.B., Seskin, E.P., 1977. Air Pollution and Human Health. The Johns Hopkins University Press, Baltimore.

Lazutka, et al., 1999. Chromosomal aberrations and sister-chromatid exchanges in Lithuanian populations: effects of occupational and environmental exposures. Mutation Research/Genetic Toxicology and Environmental Mutagenesis 445 (2), 225–239.

Lester, B.L., Seskin, E.P., 1970. Air pollution and human health. American Association for the Advancement of Science 169, 723–733.

LeTertre, A., Medina, S., Samoli, E., Forsberg, B., Michelozzi, P., Boumghar, A., Vonk, J.M., Bellini, A., Atkinson, R., Ayres, J.G., Sunyer, J., Schwartz, J., Katsouyanni, K., 2002. Short-term effects of particulate air pollution on cardiovascular diseases in eight European cities. Journal of Epidemiology and Community Health 56, 773–779.

Levy, J.I., Hammitt, J.K., Spengler, J.D., 2000. Estimating the mortality impacts of particulate matter: what can be learned from between-study variability? Environmental Health Perspective 108, 109–117.

Lewtas, J., 2007. Air pollution combustion emissions: characterization of causative agents and mechanisms associated with cancer, reproductive, and cardiovascular effects. Mutation Research/Reviews in Mutation Research 636 (1–3), 95–133.

Li, N., Sioutas, C., Cho, A., et al., 2003. Ultrafine particulate pollutants induce oxidative stress and mitochondrial damage. Environmental Health Perspectives 111, 455–460.

Liao, D., Creason, J., Shy, C., Williams, R., Watts, R., Zweidinger, R., 1999. Daily variation of particulate air pollution and poor cardiac autonomic control in the elderly. Environmental Health Perspectives 107, 521–525.

Liao, D., Duan, Y., Whitsel, E.A., Zheng, Z., Heiss, G., Chinchilli, V.M., Lin, H.M., 2004. Association of higher levels of ambient criteria pollutants with impaired cardiac autonomic control: a population-based study. American Journal of Epidemiology 159, 768–777.

Lin, C.A., Pereira, L.A., Conceicao, G.M.S., Kishi, H.S., Milani Jr., R., Braga, A.L.F., Saldiva, P.H.N., 2003. Association between air pollution and ischemic cardiovascular emergency room visits. Environmental Research 92, 57–63.

Lipfert, F.W., 1978. The Association of Human Mortality with Air Pollution: Statistical Analyses by Region, by Age, and by Cause of Death. Eureka Publications, Mantua, New Jersey.

Lipmann, M., 2007. Health effects of airborne particulate matter. The New England Journal of Medicine 357 (23), 2395–2397.

Ma, T.H., Cabrera, G.L., Chen, R., Gill, B.S., Sandhu, S.S., Valenberg, A.L., Salamone, M.F., 1994. Tradescantia micronucleus bioassay. Mutation Research 310, 221–230.

Maher, B.A., 2009. Rain and dust: magnetic records of climate and pollution. Elements 5, 229–234.

Maher, B.A., Mooreb, C., Matzka, J., 2008. Spatial variation in vehicle-derived metal pollution identified by magnetic and elemental analysis of roadside tree leaves. Atmospheric Environment 42, 364–373.

Michalska, J., Motykiewicz, G., Pendzich, J., Kalinowska, K., Midro, A., Chorazy, M., 1999. Measurement of cytogenetic endpoints in women environmentally exposed to air pollution. Mutation Research 445, 139–145.

Migliaretti, G., Cadum, E., Migliore, E., Cavallo, F., 2005. Traffic air pollution and hospital admission for asthma: a case-control approach in a Turin (Italy) population. International Archives of Occupational Environmental Health 78, 164–169.

Mitchell, R., Maher, B., Kinnersley, R., 2010. Rates of particulate pollution deposition onto leaf surfaces: temporal and inter-species magnetic analyses. Environmental Pollution 158 (5), 1472–1478.

Moller, P., et al., 2008. Air pollution, oxidative damage to DNA, and carcinogenesis. Cancer Letters 266, 84–97.

Moshammer, H., Hutter, H.P., Hauck, H., Neuberger, M., 2006. Low levels of air pollution induce changes of lung function in a panel of school children. European Respiratory Journal 27, 1138–1143.

Muzyka, V., Veimer, S., Schmidt, N., 1998. On the carcinogenic risk evaluation of diesel exhaust: benzene in airborne particles and alterations of heme metabolism in lymphocytes as markers of exposure. The Science of the Total Environment 217, 103–111.

Nielsen, P.S., Andreassen, A., Farmer, P.B., Ovrebo, S., Autrup, H., 1996a. Biomonitoring of diesel exhaust-exposed workers. DNA and haemoglobin adducts and urinary OHP as markers of exposure. Toxicology Letters 86, 27–37.

Nielsen, P.S., de Pater, N., Okkels, H., Autrup, H., 1996b. Environmental air pollution and DNA adducts in Copenhagen bus drivers—effect of GSTM1 and NAT2 genotypes on adduct levels. Carcinogenesis 17, 1021–1027.

Oberdorster, G., Oberorster, E., Oberdorster, J., 2005. Nanotoxicology: an emerging discipline evolving from studies of ultrafine particles. Environmental Health Perspectives 113, 823–839.

Pandey, S.K., Tripathi, B.D., Mishra, V.K., Prajapati, S.K., 2006. Size fractionated speciation of nitrate and sulfate aerosols in a sub-tropical industrial environment. Chemosphere 63, 49–57.

Pandey, S.K., Tripathi, B.D., Prajapati, S.K., et al., 2005. Magnetic properties of vehicles derived particulates and amelioration by *Ficus infectoria*: a keystone species. AMBIO 35, 645–647.

Park, S.K., O'Neill, M.S., Vokonas, P.S., Sparrow, D., Schwartz, J., 2005. Effects of air pollution on heart rate variability: the VA normative aging study. Environmental Health Perspectives 113, 304–309.

Pascal, M., et al., 2014. Short-term impacts of particulate matter (PM_{10}, $PM_{10-2.5}$, $PM_{2.5}$) on mortality in nine French cities. Atmospheric Environment 95, 175–184.

Peden, D.B., 2001. Air pollution in asthma: effect of pollutants on airway inflammation. Annals of Allergy, Asthma, and Immunology 87, 12–17.

Pedersen, M., et al., 2006. Cytogenetic effects in children and mothers exposed to air pollution assessed by the frequency of micronuclei and fluorescence in situ hybridization (FISH): a family pilot study in the Czech Republic. Mutation Research 608, 112–120.

Peng, R.D., Dominici, F., Pastor-Barriuso, R., et al., 2005. Seasonal analyses of air pollution and mortality in 100 US cities. American Journal of Epidemiology 161, 585–594.

Penttinen, P., Timonen, K.L., Tiittanen, P., Mirme, A., Ruuskanen, J., Pekkanen, J., 2001. Ultrafine particles in urban air and respiratory health among adult asthmatics. European Respiratory Journal 17, 428–435.

Pereira, C.E.L., et al., 2007. Ambient particulate air pollution from vehicles promotes lipid peroxidation and inflammatory responses in rat lung. Brazilian Journal Medical Biological Research 40, 1353–1359.

Peters, A., Doring, A., Wichmann, H.-E., Koenig, W., 1997. Increased plasma viscosity during an air pollution episode: a link to mortality? The Lancet 349, 1582–1587.

Peters, A., Fröhlich, M., Döring, A., Immervoll, T., Wichmann, H.-E., Hutchinson, W.L., Pepys, M.B., Koenig, W., 2001. Particulate air pollution is associated with an acute phase response in men: results from the MONICA-Augsberg study. European Heart Journal 22, 1198–1204.

Peters, J.M., Avol, E., Navidi, W., London, S.J., Gauderman, W.J., Lurmann, F., Linn, W.S., Margolis, H., Rappaport, E., Gong, H., Thomas, D.C., 1999. A study of twelve southern California communities with differing levels and types of air pollution. I: prevalence of respiratory morbidity. American Journal of Respiratory and Critical Care Medicine 159, 760–767.

Polichetti, et al., 2009. Effects of particulate matter (PM_{10}, $PM_{2.5}$ and PM_1) on the cardiovascular system. Toxicology 261, 1–8.

Pope III, C.A., 2000. Epidemiology of Fine particulate air pollution and human health: biologic mechanisms and who's at risk? Environmental Health Perspectives 108 (4), 713–723.

Pope III, C.A., Burnett, R.T., Thun, M.J., et al., 2002. Lung cancer, cardiopulmonary mortality and long-term exposure to fine particulate air pollution. Journal of American Medical Association 287, 1132–1141.

Pope III, C.A., Burnett, R.T., Thurston, G.D., Thun, M.J., Calle, E.E., Krewski, D., Godleski, J.J., 2004. Cardiovascular mortality and long-term exposure to particulate air pollution: epidemiological evidence of general pathophysiological pathways of disease. Journal of the American heart Association 109, 71–77.

Pope III, C.A., Muhlestein, J.B., et al., 2006. Ischemic heart disease events triggered by short-term exposure to fine particulate air pollution. Journal of the American Heart Association 114, 2443–2448.

Pope III, C.A., Verrier, R.L., Lovett, E.G., Larson, A.C., Raizenne, M.E., Kanner, R.E., Schwartz, J., Villegas, G.M., Gold, D.R., Dockery, D.W., 1999. Heart rate variability associated with particulate air pollution. American Heart Journal 138, 890–899.

Portney, P.R., Mullahy, J., 1990. Urban air quality and chronic respiratory disease. Regional Science and Urban Economics 20, 407–418.

Raaschou-Nielsen, O., Andersen, N.Z., Beelen, R., et al., 2013. Air pollution and lung cancer incidence in 17 European cohorts: prospective analyses from the European study of cohorts for air pollution effects(ESCAPE). Lancet Oncology 14, 813–822.

Rai, P.K., 2013. Environmental magnetic studies of particulates with special reference to biomagnetic monitoring using roadside plant leaves. Atmospheric Environment 72, 113–129.

Rai, P.K., 2015. Multifaceted health impacts of particulate matter (PM) and its management: an overview. Environmental Skeptics and Critics 4 (1), 1–26.

Rai, P.K., Panda, L.S., 2014. Dust capturing potential and air pollution tolerance index (APTI) of some road side tree vegetation in Aizawl, Mizoram, India: an Indo-Burma hot spot region. Air Quality, Atmospheric and Health 7 (1), 93–101.

Rai, P.K., Chutia, B.M., Patil, S.K., 2014. Monitoring of spatial variations of particulate matter (PM) pollution through bio-magnetic aspects of roadside plant leaves in an Indo-Burma hot spot region. Urban Forestry and Urban Greening 3, 761–770.

Rajput, M., Agrawal, M., 2005. Biomonitoring of air pollution in a seasonally dry tropical suburban area using wheat transplants. Environmental Monitoring and Assessment 101, 39–53.

Risom, L., Møller, P., Loft, S., 2005. Oxidative stress-induced DNA damage by particulate air pollution. Mutation Research 592, 119–137.

Rizzio, E., Giaveri, G., Arginelli, D., Gini, L., Profumo, A., Gallorini, M., 1999. Trace elements total content and particle sizes distribution in the air particulate matter of a rural-residential area in the north Italy investigated by instrumental neutron activation analysis. Science of the Total Environment 226, 47–56.

Rohr, C.A., Wyzga, R.E., 2012. Attributing health effects to individual particulate matter constituents. Atmospheric Environment 62, 130–152.

Rucker, R., Ibald-Mulli, A., Koenig, W., Schneider, A., Woelke, G., Cyrys, J., Heinrich, J., Marder, V., Frampton, M., Wichmann, H.E., Peters, A., 2006. Air pollution and markers of inflammation and coagulation in patients with coronary heart disease. American Journal of Respiratory and Critical Care Medicine 173, 432–441.

Saint-Georges, F., et al., 2009. Role of air pollution Particulate Matter ($PM_{2.5}$) in the occurrence of loss of heterozygosity in multiple critical regions of 3p chromosome in human epithelial lung cells (L132). Toxicology Letters 187, 172–179.

Saldiva, P.H., Clarke, R.W., Coull, B.A., Stearns, R.C., Lawrence, J., Murthy, G.G., Diaz, E., Koutrakis, P., Suth, H., Tsuda, A., Godleski, J.J., 2002. Lung inflammation induced by concentrated ambient air particles is related to particle composition. American Journal of Respiratory and Critical Care Medicine 165, 1610–1617.

Salvi, S., Blomberg, A., Rudell, B., Kelly, F., Sandstrom, T., Holgate, S.T., Frew, A., 1999. Acute inflammatory responses in the airways and peripheral blood after short-term exposure to diesel exhaust in healthy human volunteers. American Journal of Respiratory and Critical Care Medicine 159, 702–709.

Salvi, S.S., Nordenhall, C., Blomberg, A., Rudell, B., Pourazar, J., Kelly, F.J., Wilson, S., Sandstrom, T., Holgate, S.T., Frew, A.J., 2000. Acute exposure to diesel exhaust increases IL-8 and GRO-alpha production in healthy human airways. American Journal of Respiratory and Critical Care Medicine 161, 550–557.

Samet, J., Krewski, D., 2007. Health effects associated with exposure to ambient air pollution. Journal of Toxicology and Environmental Health, part A 70 (3), 227–242.

Samet, J.M., 2007. Traffic, air pollution, and health. Inhalation Toxicology 19 (12), 1021–1027.

Samet, J.M., Dominici, F., Curriero, F.C., Coursac, I., Zeger, S.L., 2000a. Fine particulate air pollution and mortality in 20 US cities, 1987–1994. New England Journal of Medicine 343 (24), 1742–1749.

Samet, J.M., Zeger, S.L., Dominici, F., Curriero, F., Coursac, I., Dockery, D.W., et al., 2000b. The National morbidity, mortality, and air pollution study. Part II: morbidity and mortality from air pollution in the United States. Research Report (Health Effects Institute) 94 (2), 5–70.

Saragnese, F., Lanci, L., Lanza, R., 2011. Nanometric-sized atmospheric particulate studied by magnetic analyses. Atmospheric Environment 45, 450–459.

Scheepers, et al., 2002. Biomarkers for occupational diesel exhaust exposure monitoring (BIOMODEM)—a study in underground mining. Toxicology Letters 134, 305–317.

Schins, R.P., Lightbody, J.H., Borm, P.J., Shi, T., Donaldson, K., Stone, V., 2004. Inflammatory effects of coarse and fine particulate matter in relation to chemical and biological constituents. Toxicology and Applied Pharmacology 195, 1–11.

Schoket, B., 1999. DNA damage in humans exposed to environmental and dietary polycyclic aromatic hydrocarbons. Mutation Research 424, 143–153.

Schwart, J., Zanobetti, A., 2005. The effect of particulate air pollution on emergency admissions for myocardial infarction: a multicity case-crossover analysis. Environmental Health Perspectives 113 (8).

Schwartz, J., 1996. Air pollution and hospital admissions for respiratory disease. Epidemiology 7, 20–28.

Schwartz, J., Dockery, D.W., 1992. Particulate air pollution and daily mortality in Steubenville, Ohio. American Journal of Epidemiology 135, 12–19.

Schwartz, J., Morris, R., 1995. Air pollution and hospital admissions for cardiovascular disease in Detroit, Michigan. American Journal Epidemiology 142, 23–35.

Schwartz, J., Park, S.K., O'Neill, M.S., Vokonas, P.S., Sparrow, D., Weiss, S.T., Kelsey, K., 2005. Glutathione-S-transferase M1, obesity, statins, and autonomic effects of particles. American Journal of Respiratory and Critical Care Medicine 172, 1529–1533.

Schwartz, J., 1993. Particulate air pollution and chronic respiratory disease. Environmental Research 62, 7–13.

Seaton, A., MacNee, W., Donaldson, K., Godden, D., 1995. Particulate air pollution and acute health effects. The Lancet 345 (8943), 176–178.

Shukla, A., Timblin, C., BeruBe, K., Gordon, T., McKinney, W., Driscoll, K., Vacek, P., Mossman, B.T., 2000. Inhaled particulate matter causes expression of nuclear factor (NF)-kB-related genes and oxidant-dependent NF-kB activation in vitro. American Journal of Respiratory Cell and Molecular Biology 23, 182–187.

Sirajuddin, M., Ravichandran, M., 2010. Ambient air quality in an urban area and its effects on plants and human beings: a case study of Tiruchiraalli, India. Kathmandu University Journal of Science, Engineering and Technology 6 (2), 13–19.

Stenfors, N., Nordenhall, C., Salvi, S.S., Mudway, I., Soderberg, M., Blomberg, A., Helleday, R., Levin, J.O., Holgate, S.T., Kelly, F.J., Frew, A.J., Sandström, T., 2004. Different airway inflammatory responses in asthmatic and healthy humans exposed to diesel. European Respiratory Journal 23, 82–86.

Sul, D., Oh, E., Im, H., Yang, M., Kim, C.W., Lee, E., 2003. DNA damage in T- and B-lymphocytes and granulocytes in emission inspection and incineration workers exposed to polycyclic aromatic hydrocarbons. Mutation Research/Genetic Toxicology and Environmental Mutagenesis 538 (1–2), 109–119.

Sun, H.L., Chou, M.C., Lue, K.H., 2006. The relationship of air pollution to ED visits for asthma differ between children and adults. American Journal of Emergency Medicine 24, 709–713.

Suwa, et al., 2002. Particulate air pollution induces progression of atherosclerosis. Journal of the American College of Cardiology 39 (6), 935–942.

Tan, W.C., Qiu, D., Liam, B.L., Ng, T.P., Lee, S.H., van Eeden, S.F., D'Yachkova, Y., Hogg, J.C., 2000. The human bone marrow response to acute air pollution caused by forest fires. American Journal of Respiratory and Critical Care Medicine 161, 1213–1217.

Taner, S., Pekey, B., Pekey, H., 2013. Fine particulate matter in the indoor air of barbeque restaurants: elemental compositions, sources and health risks. Science of the Total Environment 454-455, 79–87.

Tarantini, L., Bonzini, M., Apostoli, P., Pegoraro, V., Bollati, V., Marinelli, B., Cantone, L., Rizzo, G., Hou, L., Schwartz, J., Bertazzi, P.A., Baccarelli, A., 2009. Effects of particulate matter on genomic DNAmethylation content and iNOS promoter methylation. Environmental Health Perspectives 117 (2), 217–222.

Ulrich, M.M., Alink, G.M., Kumarathasan, P., Vincent, R., Boere, A.J., Cassee, F.R., 2002. Health effects and time course of particulate matter on the cardiopulmonary system in rats with lung inflammation. Journal of Toxicology and Environmental Health, Part A 65, 1571–1595.

Utell, M.J., Frampton, M.W., Zareba, W., et al., 2002. Cardiovascular effects associated with air pollution: potential mechanisms and methods of testing. Inhalation Toxicology 14, 1231–1247.

Valavanidis, A., Fiotakis, K., Vlachogianni, T., 2008. Airborne particulate matter and human health: toxicological assessment and importance of size and composition of particles for oxidative damage and carcinogenic mechanisms. Journal of Environmental Science and Health, Part C 26 (4), 339–362.

Veronesi, B., Oortgiesen, M., 2001. Neutogenic inflammation and particulate matter (PM) air pollution. Neurotoxicology 22, 795–810.

Vincent, R., Jumarathasan, P., Mukherjee, B., Gravel, C., Bjarnason, S., Urch, B., Speck, M., Brook, J., Tarlo, S., Zimmerman, B., Siverman, F., 2001. Exposure to urban particles ($PM_{2.5}$) causes elevation of the plasma vasopeptidesendothelin (ET)-1 and ET-3 in humans (Abstract). American Journal Respiratory and Critical Care Medicine 163, A313.

Wang, X., Hui, D., Louise, R., Xiping, X., 1997. Association between air pollution and low birth weight: a community-based study. Environmental Health Perspectives 105 (5), 514–520.

White, R.H., Spengler, J.D., Dilwali, K.M., Barry, B.E., Samet, J.M., 2005. Report of workshop on traffic, health, and infrastructure planning. Archives of Environmental and Occupational Health 60, 70–76.

Wichmann, H.E., Peters, A., 2000. Epidemiological evidence of the effects of ultrafine particle exposure. Philosophical Transaction of the Royal Society of London, Part A 358, 2751–2769.

Wilhelm, M., Eberwein, G., Holzer, J., Begerow, J., Sugiri, D., Gladtke, D., Ranft, U., 2005. Human biomonitoring of cadmium and lead exposure of child–mother pairs from Germany living in the vicinity of industrial sources (Hot Spot Study NRW). Journal of Trace Elements in Medicine and Biology 19, 83–90.

Woodruff, T., Jeanne, G., Kenneth, S., 1997. The relationship between selected causes of postneonatal infant mortality and particulate air pollution in the United States. Environmental Health Perspectives 105 (6), 608–612.

Yadav, R., Beig, G., Jaffrrey, S.N.A., 2014. The linkages of anthropogenic emissions and meteorology in the rapid increase of particulate matter at a foothill city in the Arawali range of India. Atmospheric Environment 85, 147–151.

Yang, T.H., et al., 2015. Personal exposure to particulate matter and inflammation among patients with periodontal disease. Science of the Total Environment 502, 585–589.

Yang, W., Omaye, S.T., 2009. Air pollutants, oxidative stress and human health. Mutation Research 674, 45–54.

Zanobetti, A., Schwartz, J., Dockery, D.W., 2000. Airborne particles are a risk factor for hospital admissions for heart and lung disease. Environmental Health Perspectives 108, 1071–1077.

Zeger, S.L., Dominici, F., McDermott, A., Samet, J.M., 2008. Mortality in the medicare population and chronic exposure to fine particulate air pollution in urban centers (2000–2005). Environmental Health Perspectives 116 (12), 1614–1619.

Zhao, X., Niu, J., Wang, Y., Yan, C., Wang, X., Wang, J.N., 1998. Genotoxicity and chronic health effects of automobile exhaust: a study on the traffic policemen in the city of Lanzhou. Mutation Research 415, 185–190.

CHAPTER THREE

Monitoring of Ambient Particulate Matter in South Asia with Special Reference to an Indo-Burma Hot Spot Region

3.1 AIR POLLUTION STATUS IN SOUTH ASIA

Local, regional, and global air quality issues, and regional and global environmental impacts, including climate change, should be viewed in an integrated manner (Akimoto, 2003). In the 1990s, nitrogen oxide emissions from Asia surpassed those from North America and Europe and should continue to exceed them for decades (Akimoto, 2003). Brown clouds over South Asia are primarily from biomass combustion (as the case of shifting cultivation in an Indo–Burma hot spot region) and secondarily from combustion of fossil fuels (Gustafsson et al., 2009).

Between April and November 1997, a widespread series of forest fires in Indonesia threw a blanket of thick, smoky haze over a large portion of Southeast Asia (Sastry, 2002). It led to increased global warming and hampered economic and ecological systems. Furthermore, this haze caused severe human health hazards. These examples show that biomass burning is an extremely dangerous and voluminous form of air pollution when compared to fossil fuel combustion.

Dhaka, Bangladesh, is one of the most densely populated megacities in the world, with about 15 million inhabitants within a 300-km^2 area. The city suffers from major traffic congestion from both public and private sources. The mass of particulate matter 2.5 μm ($PM_{2.5}$) is approximately 88% of the particulate matter 10 μm (PM_{10}) mass, indicating that fossil fuel is the main source of PM in Dhaka (Salam et al., 2008). A high concentration of air pollutants, such as black carbon, has already been reported in Dhaka (Salam et al., 2003, 2008; Begum et al., 2004). Vehicular emissions, biomass burning for cooking, brick kilns, and construction activities in and around the city are the main

Biomagnetic Monitoring of Particulate Matter
ISBN 978-0-12-805135-1
http://dx.doi.org/10.1016/B978-0-12-805135-1.00003-2

contributors to aerosol pollution in Dhaka (Azad and Kitada, 1998; Salam et al., 2008). The overall trace metal concentrations in Dhaka are higher than those in European (e.g., Spain, Norway) and East Asian (e.g., Taiwan) locations, but lower than those measured in Southeast Asian (Kanpur, Delhi, Mumbai, India; Lahore, Pakistan) cities (Salam et al., 2008). Mercury (Hg) is a hazardous material/metal of special concern in Asian countries (Li et al., 2009).

In Asia, it is particularly relevant to study air pollution impact in China and India in view of their large populations that include almost half of the global population. Deterioration of air quality is a problem that is directly experienced by a majority of the 300 million urban Indians, who constitute 30% of India's population (Kandlikar and Ramachandran, 2000). During the 1970s, black smoke from stacks became characteristic of Chinese industrial cities; in the 1980s, many southern cities began to suffer serious acid rain pollution; and recently, the air quality in large cities has deteriorated due to nitrous oxides (NO_x), carbon monoxide (CO), and photochemical smog, which are typical of vehicle pollution (He et al., 2002). In Beijing, about 92% of the PM in coal smoke dust is less than 10 μm (He et al., 2002), which has adverse health impacts. Similar scenarios exist in other Indian and Chinese cities. The Public Health and Air Pollution in Asia Project estimates the impact of particulate matter in Asia (Vichit-Vadakan et al., 2008). Figure 3.1 describes the status of PM pollution in different countries and their biomonitors.

3.2 INTRODUCTION TO SHIFTING CULTIVATION AND AIR QUALITY IN AN INDO-BURMA HOT SPOT REGION

In urban areas, vehicular pollution is predominant and significantly contributes to air quality related health problems (Becker et al., 2005). However, in hilly areas other factors like biomass burning through shifting cultivation for agriculture (Rai, 2012; Rai and Chutia, 2014; Rai, 2015), also exacerbate the problem of air pollution. Air quality is generally described as a combination of the physical and chemical characteristics that make air a healthful resource for human beings, animals, and plants (Joshi and Bora, 2011). Atmospheric pollutants, in both gaseous and particulate form, pose a serious threat to air quality (Bucko et al., 2011; Rai, 2013; Rai and Chutia, 2014; Rai, 2015). Most Indian cities have high concentrations of gaseous and particulate pollutants due to industrialization, badly maintained and poor roads, poor maintenance of vehicles, uses of fuels with poor environmental performance, and lack of awareness (Joshi and Chauhan, 2008; Chauhan, 2010). Nitrogen dioxide (NO_2), sulfur dioxide (SO_2), and suspended particulate matter (SPM) are regarded as major air pollutants in India (Agrawal and Singh, 2000; Rai and Chutia, 2014; Rai, 2015).

The rapid urbanization, fast, dramatic increases in the number of vehicles on the roads and other activities including soil erosion, mining, stone quarrying, and shifting cultivation in Aizawl, may lead to increases in the concentration of gaseous and particulate pollution in the ambient air. Shifting agriculture or slash-and-burn agriculture is locally called *jhooming*. Practiced since time immemorial (originating during Neolithic times), it is still the major form of agriculture in the North Eastern (NE) Himalaya of Indo-Burma hot spot region (Rai, 2012). In light of these factors, ambient air quality of Aizawl (capital of Mizoram state in NE India) has been monitored in the present study. This analysis is an attempt to investigate the air quality status and air quality index (AQI) at selected monitoring sites of Aizawl, Mizoram. At these sites, the practice during March-April onwards (summer season) is common; therefore, we also tried to investigate whether it also affects the air quality during the summer season. Until now, no research has attempted to investigate the impact of shifting cultivation on air quality.

3.3 MATERIALS AND METHODS

3.3.1 Study Area

Mizoram (21°56′–24°31′N and 92°16′–93°26′E) is one of the eight states of Northeast India and covers an area of 21,081 km². Aizawl (21°58′–21°85′N and 90°30′–90°60′E), the capital of the state is 1132 m above sea level (Figure 3.1). The Aizawl district comes under the Indo-Burma hot spot region of Northeast India (Rai, 2012, 2009). This area is of extreme ecological relevance as it comes under an Indo-Burma hot that experiences distinct seasons. The ambient air temperature normally ranges 20–30 °C in summer and 11–21 °C in winter (Laltlanchhuang, 2006).

3.3.2 Study Sites

The present study was carried out in the Aizawl district of an Indo-Burma hot spot region that was categorized into two subsites. The first study site was Ramrikawn, which is a peri-urban area including markets, bus stands, and food storage (Food Corporation of India) and the second study site was Tanhril, which is a rural area having low vehicular activity, located in the western part of Aizawl district.

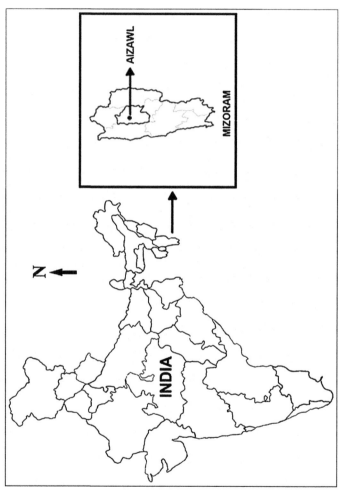

Figure 3.1 Map of the study area, Aizawl, Mizoram.

3.3.3 Ambient Air Quality Monitoring

Sampling was done using a "high volume sampler" (Envirotech APM 460) 24 h for different air pollutants (SPM, RSPM, NO_2, and SO_2) in the months of January 2012 to December 2013 with a frequency of once every a week. The apparatus was kept at a height of 2 m from the surface of the ground. At the end of the study, the samples were analyzed to determine the concentrations of different pollutants. RSPM was trapped by glass fiber filter papers (GF/A) of Whatman and SPM were collected in the separate containers at average air flow rate of $1.5\,m^3\,min^{-1}$. NO_2 and SO_2 were collected by bubbling the sample in a specific absorbing solution (sodium hydroxide for NO_2 and sodium tetrachloromercuate for SO_2) at an average flow rate of 0.2–$0.5\,min^{-1}$. The concentrations of NO_2 and SO_2 were measured by modified Jacobs–Hochheiser method (Jacob and Hochheiser, 1958) and modified West and Gaeke method (West and Gaeke, 1956), respectively. Using air pollutant data the air quality index (AQI) was calculated by modifying the following equation (Rao and Rao, 1998):

$$AQI = 1/3\,(SO_2/S_{SO_2} + NO_X/S_{NO_X} + SPM/S_{SPM}) \times 100$$

3.4 RESULTS AND DISCUSSION

Tables 3.1 and 3.2 represent the air pollutants concentration (SPM, RSPM, NO_2, and SO_2) and AQI [$(1/4\,(SO_2/S_{SO2} + NO_2/S_{NO2} + SPM/S_{SPM} + RSPM/S_{RSPM}) \times 100)$] at two monitoring sites of Aizawl. The study has shown variation in the pollutant levels during different study periods (Figures 3.2 and 3.3). The results revealed that at both the sites SPM and RSPM concentration were very high, which may have human health implications. Concentration of SPM at Ramrikawn area was 260.41, 272.26, and $100.04\,\mu g\,m^{-3}$ during winter, summer, and rainy season, respectively, where it was 157.08, 163.75, and $60.01\,\mu g\,m^{-3}$, respectively, for the same period in the Tanhril area. The standard limits as prescribed by the Central Pollution Board of India for SPM is $200\,\mu g\,m^{-3}$ for residential areas. RSPM at Ramrikawn area was 217.15, 225.13, and $59.01\,\mu g\,m^{-3}$ during winter, summer, and rainy seasons; in Tanhril area it was 122.01, 130.01, and $21.01\,\mu g\,m^{-3}$, respectively, for the same period. The standard limit prescribed by the Central Pollution Board of India for RSPM is $100\,\mu g\,m^{-3}$ for residential areas. As per the general trend, the SPM and RSPM should be high in winter season, however, we recorded their high

Table 3.1 Sources of PM pollution in different countries and their biomonitoring studies

Source of pollution	Impact	References
Coal burning station in Massachusetts	Lichen cover decreased near coal burning station	Murphy et al. (1999)
Automobile pollution in Lucknow, India	Changes in photosynthetic pigments, protein and cysteine contents; also changes in leaf area and foliar surface architecture of plants	Verma and Singh (2006)
Vehicular traffic and domestic heating in Siena (central Italy)	Acted as bioindicator and revealed amelioration of air quality during 1995.	Loppi et al. (2002)
Particulate pollution due to urban traffic and other human activities in European and North American regions	Efficient in dust capturing potential in developing and developed countries with arid and semiarid conditions	Freer-Smith et al. (2004)
Vehicular traffic and major industrial plants within the urban area of Naples	High levels of Cu, Cd, Ni, and other elements in plant leaves	Alfani et al. (2000)
Air pollution impact in Man and Biosphere Reserve, Wienerwald, Austria	Distribution of bryophytes was greatly influenced by air quality in reserve as revealed by Index of Atmospheric Purity (IAP)method	Krommer et al. (2007)
Air pollution biomonitoring (spatial and temporal variation in elemental concentration) at six sites in northern Minnesota, USA	Biomonitoring with lichens is strongly species dependent as corticolous species were more enriched in heavy metals than the terricolous species when monitored over a span of 11 years	Bennett and Wetmore (1999)
Vehicular traffic in the town of Montecatini Terme (central Italy)	As biomonitor of heavy metals	Loppi et al. (2004)
Air pollution in an urban site of Northern Italy	In relation to 29 elements studied *Hypogymnia physodes*, *Pseudevernia. furfuracea* and *Usnea gr. hirta* have a similar accumulation capacity, while that of *Parmelia sulcata* is lower	Bergamaschi et al. (2007)

Industrial complex of Cubatao, Brazil	Adverse impacts were visible injuries, reductions in net photosynthesis, growth parameters, and ascorbate concentrations, and increased F, N, and S foliar concentrations	Moraes et al. (2003)
Heavy metal pollution in Palermo city (Sicily, Italy)	*Nerium oleander* can be considered as a means of assessing dust contamination in the urban environment	Mingorance and Oliva (2006)
Global problem of carbon dioxide enrichment (climate change) on crop physiology and food quality	Adverse impacts of carbon dioxide enrichment may be decreased protein and mineral nutrient concentrations, as well as altered lipid composition in food of crop plants	DaMatta et al. (2010)
Heavy metal pollution in Hong Kong	Acted as biomonitors of metals, sulfur dioxide, and total suspended particulates	Lau and Luk (2001)
Polycyclic aromatic hydrocarbons problem in Naples, Italy	The total PAH contents in *Q. ilex* leaves were quite high	Alfani et al. (2001)
Trace metal pollution in Galicia, NW Spain	Limited efficiency as bioaccumulators of the metals studied	Aboal et al. (2004)
Problem of air pollution in industrial complex of Cubatao, SE Brazil	*P. cattleyanum and P. guajava* may be used as an accumulative indicator in tropical climates	Moraes et al. (2002)

Table 3.2 Seasonal variation of air pollutants (average of three values) and air quality index at two selected sites in Aizawl city

Site	SPM ($\mu g\,m^{-3}$)			RSPM ($\mu g\,m^{-3}$)			SO_2 ($\mu g\,m^{-3}$)			NO_2 ($\mu g\,m^{-3}$)			AQI		
	Winter	Summer	Rainy	Winter	Summer	Rainy	Winter	Summer	Rainy	Winter	Summer	Rainy	Winter	Summer	Rainy
P	260.41	272.26	100.0	217.15	225.13	59.01	4.13	5.02	1.07	22.08	23.16	10.52	94.75	98.75	30.75
R	157.08	163.75	60.01	122.01	130.01	22.01	1.04	1.15	0.68	10.07	11.13	2.04	53.25	56.25	13.5
	$200.00\,\mu g\,m^{-3}$			$100.00\,\mu g\,m^{-3}$			$80.00\,\mu g\,m^{-3}$			$80.00\,\mu g\,m^{-3}$			CPCB standard		

SPM, suspended particulate matter; RSPM, respirable suspended particulate matter; AQI, air quality index; CPCB, Central Pollution Control Board, New Delhi, India; P, Peri-urban area; R, Rural area.

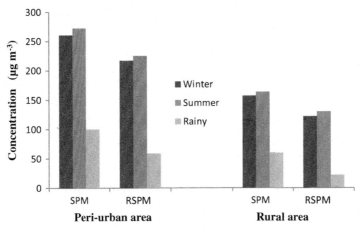

Figure 3.2 Seasonal variation of SPM and RSPM at both study sites.

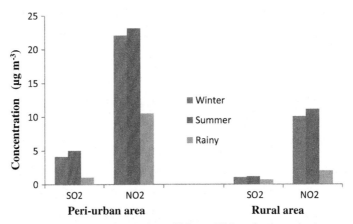

Figure 3.3 Seasonal variation of SO_2 and NO_2 at both study sites.

values during the summer season, which may be due to the impact of biomass burning through shifting cultivation. Biomass burning may lead to increased emission of particulates in the air. The highest concentration ($5.02\,\mu\mathrm{g\,m^{-3}}$) of SO_2 was also recorded during the summer season at Ramrikawn area, which was again 79.28% higher when compared with the Tanhril area. Similarly, the highest concentration ($23.16\,\mu\mathrm{g\,m^{-3}}$) of NO_2 was recorded during the summer season at Ramrikawn area, which was 51.98% higher when compared with the Tanhril area (Rai and Chutia, 2014; Rai, 2015).

In the present study, the amount of SPM and RSPM at the peri-urban area was much higher than the prescribed limits of the Central Pollution Control Board of India, while the concentration of NO_2 and SO_2 was still under the prescribed limits (Rai and Chutia, 2014). The air quality index is one of the important tools available for analyzing and representing air quality status uniformly. The concentrations of the major pollutants are monitored and subsequently converted into AQI (Table 3.1) and using a standard formula, a rating scale was also calculated (Table 3.2). The higher value of an index refers to a higher level of air pollution (Chauhan et al., 2010). In the present investigation, the SPM, RSPM, NO_2, and SO_2 levels at both selected sites were used to calculate AQI. The AQI values at the peri-urban area were 94.75, 98.75, and 30.75 during winter (heavy air pollution; HAP), summer (HAP), and rainy seasons (light air pollution; LAP). In the rural area, AQI ranged from moderate air pollution during summer and winter to clean air during rainy seasons (Table 3.3).

Meteorological factors play an important role in air pollution (Chauhan et al., 2010; Shukla, Dalal, and Chaudhry, 2010). In the above study, the concentrations of SPM, RSPM, NO_2, and SO_2 were observed at maximum in summer season compared to the winter and rainy seasons. As discussed earlier, the problem of shifting cultivation may be responsible for higher values of air quality parameters during the summer season. During the winter season there is increased atmospheric stability, which in turn allows for less general circulation and thus more stagnant air masses. This prevents an upward movement of air, hence atmospheric mixing is retarded and pollutants are trapped near the ground. Secondly, cold starts in winter lead to

Table 3.3 Rating of Air Quality Index Values at Different Study Sites

Index Values	Remarks	Peri-Urban Area			Rural Area		
		Winter	Summer	Rainy	Winter	Summer	Rainy
0–25	Clean air (CA)	HAP	HAP	LAP	MAP	MAP	CA
26–50	Light air pollution (LAP)						
51–75	Moderate air pollution (MAP)						
76–100	Heavy air pollution (HAP)						
>100	Severe air pollution (SAP)						

longer periods of incomplete combustion and longer warm-up times for catalytic converters, which generate more pollution (Shukla et al., 2010; Faiz et al., 1995; Rai, 2015).

3.5 CONCLUSIONS

Pollution levels of the two sites in an Indo–Burma hot spot region revealed that the values of air quality parameters were highest during the summer season due to the impact of shifting cultivation. The amount of SPM and RSPM at a peri-urban area (Ramrikawn) was much higher than the prescribed limits of CPCB (1994) of India, while the concentration of NO_2 and SO_2 was still under the prescribed limits. High values of SPM and RSPM may have human health implications in this region of ecological relevance. The AQI values were found to be higher at Ramrikawn during the summer and winter seasons, thus rating it as a heavy air pollution zone. The increased level of air pollution at Ramrikawn is mainly due to higher vehicular emissions combined with the impact of shifting cultivation particularly during the summer season.

REFERENCES

Aboal, J.R., Fernandez, J.A., Carballeira, A., 2004. Oak leaves and pine needles as biomonitors of airborne trace elements pollution. Environmental and Experimental Botany 51, 215–225.

Agrawal, M., Singh, J., 2000. Impact of coal power plant emission on the foliar elemental concentrations in plants in a low rainfall tropical region. Environmental Monitoring and Assessment 60, 261–282.

Akimoto, H., 2003. Global air quality and pollution. Science 302, 1716–1719.

Alfani, A., Baldantoni, T., Maisto, G., Bartoli, G., Virzo De Santo, A., 2000. Temporal and spatial variation in C, N, S and trace element contents in the leaves of *Quercus ilex* within the urban area of Naples. Environmental Pollution 109, 119–129.

Alfani, A., Maisto, G., Prati, M.V., Baldantoni, D., 2001. Leaves of *Quercus ilex* L. as biomonitors of PAHs in the air of Naples (Italy). Atmospheric Environment 35, 3553–3559.

Azad, A.K., Kitada, T., 1998. Characteristics of the air pollution in the city of Dhaka, Bangladesh in winter. Atmospheric Environment 32, 1991–2005.

Becker, S., Dailey, L.A., Soukup, J.M., Grambow, S.C., Deulin, R.B., Hunary, Y.C., 2005. Seasonal variation in air pollution particle–induced inflammatory mediator release and oxidative stress. Environmental Health Perspectives 113, 1032–1038.

Begum, B.A., Kim, E., Biswas, S.K., Hopke, P.K., 2004. Investigation of sources of atmospheric aerosol at urban and semi-urban areas in Bangladesh. Atmospheric Environment 38, 3025–3038.

Bennett, J.P., Wetmore, C.M., 1999. Changes in element contents of selected lichens over 11 years in northern Minnesota. USA Environmental and Experimental Botany 41, 75–82.

Bergamaschi, L., Rizzio, E., Giaveri, G., Loppi, S., Gallorini, M., 2007. Comparison between the accumulation capacity of four lichen species transplanted to an urban site. Environmental Pollution 148, 468–476.

Bucko, M.S., Magiera, T., Johanson, B., Petrovsky, E., Pesonen, L.J., 2011. Identification of magnetic particulates in road dust accumulated on roadside snow using magnetic, geochemical and micro-morphological analysis. Environmental Pollution 159, 1266–1276.

Chauhan, A., 2010. Photosynthetic pigment changes in some selected trees induced by automobile exhaust in Dehradun, Uttarakhand. New York Science Journal 3 (2), 45–51.

Chauhan, A., Pawar, M., Kumar, R., Joshi, P.C., 2010. Ambient air quality status in Uttarakhand (India): case study of Haridwar and Dehradun using air quality index. Journal of American Science 6 (9), 565–574.

CPCB, 1994. National Ambient Air Quality Standards. Central Pollution Control Board, Gazette Notification, New Delhi.

DaMatta, F.M., Grandis, A., Arenque, B.C., Buckeridge, M.S., 2010. Impacts of climate changes on crop physiology and food quality. Food Research International 43, 1814–1823.

Faiz, A., Surhid, G., Emaad, B., 1995. Air pollution from motor vehicles: issues and options from latin American countries. The science of the total environment 169, 303–310.

Freer-Smith, P.H., El-Khatib, A.A., Taylor, G., 2004. Capture of particulate pollution by trees: a comparison of species typical of semi-arid areas (*Ficus nitida* and *Eucalyptus globulus*) with European and North-American species. Water, Air, and Soil Pollution 155, 173–187.

Gustafsson, et al., 2009. Brown clouds over south Asia: biomass or fossil fuel combustion? Science 323, 495–498.

He, K., Huo, H., Zhang, Q., 2002. Urban air pollution in China: current status, characteristics and progress. Annual Review Energy Environ 27, 397–431.

Jacob, M.B., Hochheiser, S., 1958. Continuous sampling and ultra-micro determination of nitrogen dioxide in air. Analytical Chemistry 32, 426.

Joshi, N., Bora, M., 2011. Impact of air quality on physiological attributes of certain plants. Report and Opinion 3 (2), 42–47.

Joshi, P.C., Chauhan, A., 2008. Performance of locally grown rice plants (*Oryza sativa* L.) exposed to air pollutants in a rapidly growing industrial area of district Haridwar, Uttarakhand. India Life Science Journal 5 (3), 41–45.

Kandlikar, M., Ramachandran, G., 2000. The causes and consequences of particulate air pollution in urban India: a synthesis of the Science. Annual Review of Energy and the Environment 26, 29–84.

Krommer, V., Zechmeister, H.G., Roder, I., Scharf, S., Hanus-Illnar, A., 2007. Monitoring atmospheric pollutants in the biosphere reserve Wienerwald by a combined approach of biomonitoring methods and technical measurements. Chemosphere 67, 1956–1966.

Laltlanchhuang, S.K., 2006. Studies of the Impact of Disturbance on Secondary Productivity of Forest Ecosystem with Special Reference to Surface, Sub-surface Litter Insect and Other Non- Insect Groups (M.Sc. dissertation). Mizoram University.

Lau, O.W., Luk, S.F., 2001. Leaves of *Bauhinia blakeana* as indicators of atmospheric pollution in Hong Kong. Atmospheric Environment 35, 3113–3120.

Li, P., Feng, X.B., Qiu, G.L., Shang, L.H., Li, Z.G., 2009. Mercury pollution in Asia: a review of the contaminated sites. Journal of Hazardous Materials 168, 591–601.

Loppi, S., Frati, L., Paoli, L., Bigagli, V., Rossetti, C., Bruscoli, C., Corsini, A., 2004. Biodiversity of epiphytic lichens and heavy metal contents of *Flavoparmelia caperata* thalli as indicators of temporal variations of air pollution in the town of Montecatini Terme (central Italy). Science of the Total Environment 326, 113–122.

Loppi, S., Ivanov, D., Boccardi, R., 2002. Biodiversity of epiphytic lichens and air pollution in the town of Siena (central Italy). Environmental Pollution 116, 123–128.

Mingorance, M.D., Oliva, S.R., 2006. Heavy metal content in *N. Oleander* leaves as urban pollution assessment. Environmental Monitoring and Assessment 119, 57–68.

Moraes, R.M., Delitti, W.B.C., Moraes, J.A.P.V., 2003. Gas exchange, growth, and chemical parameters in a native Atlantic forest tree species in polluted areas of Cubatao, Brazil. Ecotoxicology and Environmental Safety 54, 339–345.

Moraes, R.M., Klumpp, A., Furlan, C.M., Klumpp, G., Domingos, M., Rinaldi, M.C.S., Modesto, I.F., 2002. Tropical fruit trees as bioindicators of industrial air pollution in southeast Brazil. Environment International 28, 367–374.

Murphy, K.J., Alpert, P., Cosentino, D., 1999. Local impacts of a rural coal-burning generating station on lichen abundance in a New England forest. Environmental Pollution 105, 349–354.

Rai, P.K., 2015. Environmental Issues and Sustainable Development of North East India. Lambert Academic Publisher, Germany, p. 317.

Rai, P.K., 2009. Comparative assessment of soil properties after bamboo flowering and death in a tropical forest of Indo-Burma hot spot. AMBIO: A Journal on Human Environment 38 (2), 118–120.

Rai, P.K., 2012. Assessment of multifaceted environmental issues and model development of an Indo-Burma hot spot region. Environmental Monitoring and Assessment 184, 113–131.

Rai, P.K., 2013. Environmental magnetic studies of particulates with special reference to biomagnetic monitoring using roadside plant leaves. Atmospheric Environment 72, 113–129.

Rai, P.K., Chutia, B.M., 2014. Assessment of ambient air quality status before and after shifting cultivation in an Indo-Burma hot spot region. International Research Journal of Environmental Sciences 3 (11), 1–5.

Rao, M.N., Rao, H.N.V., 1998. Air Pollution. Tata McGraw Hill Publishing Company Limited, New Delhi.

Salam, A., Bauer, H., Kassin, K., Ullah, S.M., Puxbaum, H., 2003. Aerosol chemical characteristics of a mega-city in Southeast Asia (Dhaka, Bangladesh). Atmospheric Environment 37, 2517–2528.

Salam, et al., 2008. Characteristics of atmospheric trace gases, particulate matter, and heavy metal pollution in Dhaka, Bangladesh. Air Quality, Atmosphere and Health 1, 101–109.

Sastry, N., 2002. Forest fires, air pollution, and mortality in southeast Asia. Demography 39 (1), 1–23.

Shukla, V., Dalal, P., Chaudhry, D., 2010. Impact of vehicular exhaust on ambient air quality of Rohtak city, India. Journal of Environmental Biology 31 (6), 929–932.

Verma, A., Singh, S.N., 2006. Biochemical and ultra structural changes in plant foliage exposed to auto pollution. Environmental Monitoring and Assessment 120, 585–602.

Vichit-Vadakan, A., Vajanapoom, N., Ostro, B., 2008. The public health and air pollution in Asia (PAPA) project: estimating the mortality effects of particulate matter in Bangkok, Thailand. Environmental Health Perspectives 116 (9), 1179–1182.

West, P.W., Gaeke, G.C., 1956. Fixation of sulphur dioxide as sulfitomercurate (II) and subsequent colorimetric determination. Analytical Chemistry 28, 1816–1819.

CHAPTER FOUR

Management Approaches of Particulate Matter: Existing Technologies and Advantages of Biomagnetic Monitoring Methodology

4.1 INTRODUCTION

The beginning of this chapter briefly discusses the management options associated with particulate matter (PM) pollution. The persistence of air pollutants and the rapid rise in vehicle travel in recent decades have raised concerns within the planning sector (Stone et al., 2007; Rai, 2015). Harrison and Yin (2000) in their review supported that a single major or trace components of the PM are responsible for the adverse health effects. Costa (2004) described the interaction among PM constituents in the light of scientific and regulatory agendas. Ugwuanyi and Obi (2002) used the environmental impact matrices of the patients versus diseases and found that pollution is already affecting the quality of life and productivity of the people (particularly farmers) in Benue state of Nigeria. Due to extreme health impacts of PM, proper caution should be taken in setting air quality standards (McClellan, 2002).

Further, comprehensive understanding about the health consequences of exposure to traffic is necessary for formulating mitigation policies (Samet and Krewski, 2007). Wilson et al. (2005) emphasized the need of intraurban assessment of exposure to PM pollution. Epidemiological meta-studies for the health impacts of fine PM may be used to predict the number of premature deaths and some morbidity impacts from prevailing ambient concentrations (Pearce and Crowards, 1996; Rai, 2015).

Air quality modeling may be a useful approach for the management of air quality. Tchepel et al. (2007) attempted a modeling approach to estimate population exposure to benzene through inhalation and suggested

Biomagnetic Monitoring of Particulate Matter
ISBN 978-0-12-805135-1
http://dx.doi.org/10.1016/B978-0-12-805135-1.00004-4

that the aforesaid model may be used in combination with human bio-monitoring in order to select who should be monitored and where the monitoring should be done, as well as for interpretation and extrapolation of biomonitoring results. Stocker (2000) suggested a methodology for evaluating air quality impacts on meteorological model to simulate the meteorological conditions that correlate with prescribed fire and wildfire activity in Colorado. According to Stocker (2000), meteorological fields are input into an air quality model that simulates transport and secondary aerosol formation for certain pollutants. Also, air quality indices may be helpful to assess the human health implications (Dimitriou et al., 2013; Rai, 2015).

Specific model systems may be developed in order to estimate the concentration of air pollutants to which humans are exposed (Borrego et al., 2009; Rai, 2015). The operational air quality forecasting modeling system is composed by the chemistry-transport model CHIMERE, forced by the MM5 meteorological fields (Monteiro et al., 2005; Borrego et al., 2009; Rai, 2015). Oxidative and immune effects of PM pollution have been demonstrated in several in vitro and animal models within European projects (Sandstrom et al., 2005). Abbas et al. (2009) extensively studied PM-induced gene expression of volatile organic carbons and polyaromatic hydrocarbons metabolizing enzymes in an in vitro coculture lung model, which may be a powerful tool to identify the mechanisms by which air pollution PM induces adverse health effects.

Vegetation is used to filter the PM concentration and consequently provide health benefits (Tiwary et al., 2007; Rai, 2015) (Table 4.1). Bio-monitoring with screening of potent plants may be emphasized in the regulation of air quality. Wolterbeek and Verburg (2004) observed correlations between moss metal concentrations and epidemiological data on health and mortality rates in the Netherlands, and their data suggested that correlation studies between biomonitoring data on metal air pollution and (epidemiological) health data may prove valuable in turning attention to specific metal-health issues and in directing further study into possible dose–response mechanisms in air-associated metal epidemiology. Okona-Mensah et al. (2005) described in his review regarding novel biomonitoring approach to evaluate exposure, uptake, and the role of high potency polycyclic aromatic hydrocarbons (PAHs) in air pollution-related lung cancers.

Education and awareness of people toward the adverse health impacts of PM pollution may be very useful in its management. Dorevitch et al. (2008)

Table 4.1 List of potent biomonitors and bioindicators under different sources of particulate matter pollution

Source of pollution	Plants as biomonitors	Impact	References
Coal burning station in Massachusetts	Lichens on *Acer saccharinum* and *Populus deltoides*	Lichen cover decreased near coal burning station	Murphy et al. (1999)
Automobile pollution in Lucknow, India	*Ficus religiosa* and *Thevetia nerifolia*	Changes in photosynthetic pigments, proteins, and cysteine content; also changes in leaf area and foliar surface architecture of plants	Singh and Verma (2007)
Vehicular traffic and domestic heating in Siena (central Italy)	57 species of lichens	Acted as bioindicator and revealed amelioration of air quality over 1995	Loppi et al. (2002)
Particulate pollution due to urban traffic and other human activities in European and North American regions	*Ficus nitida* and *Eucalyptus globules, Quercus petraea, Alnus glutinosa, Fraxinus excelsior, Acer pseudo-platanu, Pseudotsuga menziesii*	Efficient in dust capturing potential in developing and developed countries with arid and semiarid conditions	Freer-Smith et al. (2004)
Vehicular traffic and major industrial plants within the urban area of Naples (Italy)	*Quercus ilex*	High levels of Cu, Cd, Ni, and other elements in plant leaves	Alfani et al. (2000)
Air pollution impact in man and in the biosphere reserve Wienerwald, Austria	Bryophytes: *Scleropodium purum, Hypnum cupressiforme* and *Abietinella abietina*	Distribution of bryophytes was greatly influenced by air quality in reserve as revealed by Index of Atmospheric Purity method	Krommer et al. (2007)
Air pollution biomonitoring (spatial and temporal variation in elemental concentration) at six sites in northern Minnesota, USA	Lichens: *Cladina rangiferina, Evernia mesomorpha, Hypogymnia physodes,* and *Parmelia sulcata*	Biomonitoring with lichens is strongly species dependent as corticolous species were more enriched in heavy metals than the terricolous species when monitored over a span of 11 years	Bennett and Wetmore (1999)
Vehicular traffic in the town of Montecatini Terme (central Italy)	*Flavoparmelia Caperata*	As biomonitor of heavy metals	Loppi et al. (2004)

Continued

Table 4.1 List of potent biomonitors and bioindicators under different sources of particulate matter pollution—Cont'd

Source of pollution	Plants as biomonitors	Impact	References
Air pollution in an urban site of northern Italy	Lichens: *H. physodes, P. sulcata, Pseudevernia furfuracea* and *Usnea gr. hirta*	In relation to 29 elements studied, *H. physodes, P. furfuracea* and *U. gr. hirta* have a similar accumulation capacity, while that of *P. sulcata* is lower	Bergamaschi et al. (2007)
Industrial complex of Cubatao, Brazil	*Tibouchina pulchra*	Adverse impacts were visible injuries, reductions in net photosynthesis, growth parameters, and ascorbate concentrations, and increased F, N, and S foliar concentrations	Moraes et al. (2003)
Heavy metal pollution in Palermo city (Sicily, Italy)	*Nerium oleander* L.	*N. oleander* can be considered as a means of assessing dust contamination in the urban environment	Mingorance and Oliva (2006)
Global problem of carbon dioxide enrichment (climate change) on crop physiology and food quality	Crop plants	Adverse impacts of carbon dioxide enrichment may be decreased protein and mineral nutrient concentrations, as well as altered lipid composition in food of crop plants	DaMatta et al. (2010)
Heavy metal pollution in Hong Kong	*Bauhinia blakeana*	Acted as biomonitors of metals, sulfur dioxide, and total suspended particulates	Lau and Luk (2001)
Polycyclic aromatic hydrocarbons (PAH) problem in Naples, Italy	*Q. ilex*	The total PAH content in *Q. ilex* leaves was quite high	Alfani et al. (2001)
Trace metal pollution in Galicia, NW Spain	*Quercus robur* leaves and *Pinus pinaster* needle	Limited efficiency as bioaccumulators of the metals studied	Aboal et al. (2004)
Problem of air pollution in industrial complex of Cubatao, SE Brazil	*Psidium guajava, Psidium cattleyanum* and *Mangifera indica*	*P. cattleyanum* and *P. guajava* may be used as an accumulative indicator in tropical climates	Moraes et al. (2002)

assessed the efficacy of an outdoor air pollution program in a community at risk for asthma, and their findings after an exhaustive survey recommended that air quality education efforts should be further developed, evaluated, and promoted for the general public, for people with underlying cardiopulmonary disease, and given the documented health disparities within the general population, for low-income and minority communities.

El-Fadel and Massoud (2000) also emphasized the need for health-based economic assessment originating due to PM pollution in urban areas. Regarding socioeconomic and related components, the benefits and costs associated with Canada-Wide Standards for PM are highly uncertain and controversial (Adamowicz et al., 2004). Likewise, similar recommendations of cost/benefit analysis of remedial measures including an assessment of the impact of urban air quality on human health was submitted by Mitchell et al. (2000). Epidemiology-based exposure–response functions were used in order to assess the economics of health impacts in Shanghai, China, and it was estimated that the total economic cost of health impacts due to PM pollution in urban areas of Shanghai in 2001 was approximately 625.40 million US dollars (Kan and Chen, 2004). The health benefits of pollution reduction are compared with the investment costs for the new strategies in Shanghai, China, and the results showed a benefit-to-cost ratio in the range of 1–5 for the power-sector initiative and 2–15 for the industrial-sector initiative (Li et al., 2003). This study by Li et al. (2003) provides economic grounds for supporting investments in air pollution control in developing cities of Asia.

Airborne PMs are associated with increased mortality, and estimates have been used to forecast the impact on life expectancy (Coyle et al., 2003). Sultan (2007) emphasized the role that buildings' ventilation and filtration systems play in their evaluation of the health risks from outdoor PM pollution. Modern tools like satellite remote sensing may be applied in monitoring of PM (Gupta et al., 2006).

Therefore, in the light of the previous discussion, a well-defined particulate pollution control policy structure is the need of the hour in view of pollution's adverse impacts on flora and fauna, including human beings (Rai, 2013). Craig et al. (2008) prepared a guidance document in order to reflect critical science and policy aspects of air quality risk management including (1) health effects, (2) air quality emissions, measurement and modeling, (3) air quality management interventions, and (4) clean air policy challenges and opportunities. It was based on findings of five annual meetings of the Network for Environmental Risk Assessment and Management International Colloquium Series on Air Quality Management (2001–2006) as well as other research of international repute (Rai, 2013, 2015).

4.2 FEASIBILITY OF MAGNETIC BIOMONITORING APPROACH OF PARTICULATES FOR ROADSIDE PLANT LEAVES

In the light of several health hazard concerns mentioned in Chapter 2 of this book, it is necessary to have a general understanding of the existing technologies, their limitations, and the biomonitoring potential of plant leaves (Rai, 2013).

4.2.1 Existing Technologies: Limitations

Dust particles can be removed from the atmosphere by dry, wet, or occult deposition (National Expert Group on Transboundary Air Pollution, 2001). Dry deposition is the removal of pollutants by sedimentation under gravity, diffusion processes (i.e., Brownian motion), or by turbulent transfer resulting in impaction and interception (Beckett et al., 1998). However, existing technologies for abatement of particulates are not cost-effective. Okona-Mensah et al. (2005) mentioned the use of benzo[a]pyrene (B[a]P), dibenzo[a, h]anthracene (DBA), and dibenzo[a, l]pyrene (DB[a, l]P) in control of PAHs, however, hand in hand they reviewed the advantages of a biomonitoring approach over these chemicals as they are carcinogenic. It is well known that vehicle-derived PM10 values decrease not only with increased distance from roads but also with increased height (e.g., Maher et al., 2008; Mitchell and Maher, 2009; Mitchell et al., 2010; Rai, 2013). Therefore, conventional monitoring stations are not an option for particulate pollution monitoring as they are distantly located from residential areas and their height is in excess of 3 m (Mitchell et al., 2010).

Several volatile and semivolatile organic compounds are also frequently analyzed using passive biomonitors (Eriksson et al., 1989; Simonich and Hites, 1994; Davidson et al., 2003; Urbat et al., 2004), however, vegetation-atmosphere partitioning seems to be effective in abatement approaches (Calamari et al., 1991; Yang et al., 1991; Jensen et al., 1992; Franzaring, 1997;

Ockenden et al., 1998; Kylin et al., 2002; Davidson et al., 2003; Urbat et al., 2004; Rai, 2011a,b, 2013).

4.2.2 Biomonitoring and Biomagnetic Monitoring of Particulates: Advantages

Although there are many conventional (physical and chemical) devices for assessment of air pollution, biomonitoring is an efficient tool (Rai, 2011a,b). Biological monitors are organisms that provide quantitative information on some aspects of their environment, such as how much of a pollutant is present (Martin and Coughtrey, 1982). In this regard, the air cleansing capacity of urban trees presents an alternative approach to foster an integrated approach to the sustainable management of urban ecosystems (Nowak et al., 2002, 2006; Randolph, 2004; Rai, 2013).

Biomonitoring of dust pollution and its biochemical impact has attracted the attention of both national and international scientific communities. Internationally, the quantification and valuation of pertinent ecosystem services have drawn much attention over the last decade (Nowak, 1994; Taha, 1996, 1997; Beckett et al., 1998; McPherson et al., 1997, 1998, 1999; Nowak et al., 1998, 2000, 2002, 2006; Rosenfeld et al., 1998; Scott et al., 1998; Akbari et al., 2001; Akbari, 2002; Rai, 2013; Janhall, 2015; Sgrigna et al., 2015).

Besides the application of advanced technologies in prevention and amelioration, the option of employing natural mechanisms to effect ecological abatement is occasionally adopted by design but commonly contributes by default. The main advantage of using plants as biomonitors is that they are widespread, providing a high density of sampling points (Moreno et al., 2003). Moreover, the most economical and reasonable method for biomonitoring heavy metal levels in the atmosphere is by using plants (Çelik et al., 2005). Among trees, evergreen species are better traps for particles than deciduous ones because of their greater leaf longevity, which can accumulate pollutants throughout the year (Gratani and Varone, 2006, 2007). Furthermore, it might be worth mentioning that conifers also trap better because of larger leaf area.

Therefore, vegetation is an efficient sink for dust originating from diverse sources (Fowler et al., 1989; Rai, 2013; Janhall, 2015; Sgrigna et al., 2015). Dust particles from the air mainly adhere to the outside of plants. This is in contrast to air-polluting gases and very small particles (<0.1 μm), which are largely absorbed via an important part via the stomata into the leaves (Fowler et al., 1989). The use of different plant materials as biomonitors of anthropogenic contamination is discussed in detail in Markert (1993).

Likewise, in the past decade in many parts of the world there has been increased interest in the study of tree leaves as bioaccumulators of trace elements/metals present in dust, in the surroundings of industrial facilities (Helmisaari et al., 1995; Nieminen and Helmisaari, 1996; Bussotti et al., 1997; Giertych et al., 1997; Mièieta and Murín, 1998; Rautio et al., 1998a,b,c) and in urban environments (Alfani et al., 1996; Monaci et al., 2000), although few studies have been made of rural and background areas (Loppi et al., 1997; Ukonmaanaho et al., 1998). Lehndorff and Schwark (2009) investigated the spatial distribution of three-ring polycyclic aromatic hydrocarbons and their derivatives (PAH-3) in the Greater Cologne Conurbation using pine needle as passive samplers.

Lichens, bryophytes, or mosses, and certain conifers have recently been proven to be potent biomonitoring tools of air pollution (Al-Alawi et al., 2007; Al-Alawi, and Mandiwana, 2007; Larsen et al., 2007; Tretiach et al., 2007; Nali et al., 2007; Batarseh et al., 2008). Plants as well as lichens have also been used in an integrated way for diagnosis of air quality (Nali et al., 2007). Use of pine as well as cypress bark was found fruitful in biomonitoring of air pollutants, particularly heavy metals (Al-Alawi et al., 2007; Al-Alawi, and Mandiwana, 2007; Batarseh et al., 2008; Rai, 2013).

However, biomagnetic monitoring, using tree leaves as sampling surfaces, can generate high spatial-resolution PM_{10} proxy data (Hansard et al., 2011). Since atmospheric pollutants also consist of a complex mixture of magnetic particles, which are derived from iron impurities in the fuel, biomagnetic monitoring through roadside plant leaves is extremely relevant in the present scenario (Hansard et al., 2011). Also, tree leaves are abundant in number and hence are convenient for sampling. Lichens and mosses may be less abundant in severely polluted urban areas and they are also climate specific unlike roadside tree plant leaves. Therefore, in urban areas higher plants are mostly suitable for monitoring dust pollution as lichens and mosses are often missing (Al-Alawi and Mandiwana, 2007). Moreover, magnetic techniques, using natural surfaces as passive collectors of particulate pollution, are sensitive, rapid, and relatively cheap (Mitchell et al., 2010; Rai, 2013). In Chapter 5, we will discuss and review biomagnetic monitoring in detail.

4.3 METHODOLOGY: BIOMONITORING THROUGH MAGNETIC PROPERTIES

Each sample used for magnetic biomonitoring comprises six to eight leaves, sampled from the outer canopy at breast height (1.5 m) and preference is usually given to the oldest leaves from the newest twig in order to select leaves of similar age and exposure time (Matzka and Maher, 1999; Maher et al., 2008). The surface areas of the leaves are either calculated by using graph paper, leaf area meter, or by digitizing their computer-scanned images. For each sample, six to eight leaves are packed into 8/10 cm³ plastic sample holders for magnetic measurements. All leaf samples may be refrigerated at 5 °C before being analyzed.

For geomagnetic studies, leaf samples are magnetized (at room temperature) with incremental, pulsed direct current (DC) fields of 20, 50, 100, and 300 milli Tesla (mT) and 1 Tesla (T), using a magnetometer (Molspin pulse magnetizer). The resultant isothermal (i.e., room temperature) magnetic remanences (IRMs) are generally measured using a cryogenic/other applicable magnetometer. Figure 5.1 in Chapter 5 explains the concise methodology for studying the magnetic properties of tree leaves.

In India, the Indian Institute of Geomagnetism (IIGM) and its centers provide different categories of magnetometers (Molspin MINISPIN Spinner, AGICO JR-6 high-sensitivity dual speed spinner and Molspin vibrating sample magnetometer for hysteresis loop) in order to investigate the different phenomena linked with environmental geomagnetism.

The deposition of PM particles on plant leaf surfaces has been demonstrated to result in measurable magnetic properties, including magnetic remanence (Maher et al., 2008). Magnetic remanence and magnetic susceptibility are the important terms that are frequently used in the study of magnetic properties. Magnetic remanence is the magnetization remaining after a sample has been placed in and then removed from an applied DC field. Magnetic susceptibility is defined as the ratio of magnetization induced to the intensity of magnetizing field. Two threshold values of low field magnetic susceptibility (χ_{LF}) and frequency dependent susceptibility percentage ($\chi_{FD}\%$) discriminate ferromagnetic minerals of these sizes and can act as a tracer of magnetic pollution (Blundell et al., 2009). Magnetic susceptibility measurements are widely used to map and monitor the heavy metal pollution of soils (Blundell et al., 2009).

Anhysteretic remanence (ARM), incremental acquisition of magnetic remanence (IRM), and high-field IRM (HIRM) are some of the other magnetic parameters used in the field of biomagnetic monitoring.

Magnetic parameters, their monitoring, and assessment are also described in Walden et al. (1999). Like magnetic remanance and magnetic susceptibility, another important parameter is isothermal magnetic remanence values, which are usually calculated by using a cryogenic magnetometer while magnetic susceptibility is measured with a single sample susceptibility sensor (Bartington Instruments Ltd). Each IRM value is determined by taking the mean of three acquisition–measurement cycles (McIntosh et al., 2007). Since the measurements are also nondestructive, all parameters can be determined from a single sample. The leaves are rinsed in a HNO_3 solution for two to three days and then their remanence at 1 T may be remeasured in order to identify the proportion of the saturation remanence (SIRM) removed by the digestion procedure (Maher et al., 2008).

Two-dimensional (2D) magnetization (i.e., the magnetic moment per leaf area) acts as an important parameter for roadside plant leaves (Matzka and Maher, 1999; Pandey et al., 2005). It is measured in units of amperes ($A = Am^2/m^2$). After measurement, a small number of leaves are cleaned with water or detergent to determine their background magnetization (Matzka and Maher, 1999). Table 4.2 provides a descriptive account of different magnetic parameters used in magnetic studies of particulates derived from varying natural as well as anthropogenic sources and their assessment.

Statistical tools may be applied in order to assess the linkages between different magnetic parameters and biochemical parameters of plants. Correlations between magnetic parameters, PM, and elemental concentrations (heavy metals like Fe or Pb) of tree leaves may be further investigated using linear regression and t-tests (Maher et al., 2008). The 2D-magnetization values of plant leaves may be correlated with the dust content and other constituents of dust (e.g., heavy metals), hence, acting as an indicator/proxy of particulate pollution.

Zhang et al. (2007) demonstrated that χARM can be used to normalize for particle size effects as efficiently as common reference elements such as Al while investigating the role of magnetic studies normalizing heavy metal concentrations for particle size effects in intertidal sediments in the Yangtze Estuary, China.

Having described the existing management technologies and the advantages of biomagnetic monitoring methodology, we will review the concept of biomagnetic monitoring and its multifaceted impacts in Chapter 5.

Table 4.2 List of different magnetic parameters, related instrumentation, and units

S. No.	Parameters	Explanation	Instruments used	Unit
1.	K or χ	Magnetic susceptibility: Defined as ratio of magnetization induced to intensity of magnetizing field; usually proportional to concentration of ferromagnetic (γFe_2O_3); generally measured on a volume (K) or mass specific basis (χ); *The bulk magnetic susceptibility measures the magnetizability of a material and is dominated, but not solely carried by ferromagnetic minerals like iron oxides.*	Single sample susceptibility sensor (Bartington Instruments Ltd); AGICO, Czech Republic; AGICO Kappabridge KLY-2; AGICO KLY-3S Kappabridge (operating at 875 Hz frequency and $300 Am^{-1}$ RMS field intensity)	K (dimensionless); χ ($m^3 kg^{-1}$); m for plant leaves
2.	χ_{ARM} ARM anhysteretic remanent magnetization	Sample acquires anhysteretic remanence when subjected to a decreasing alternating field; respond to both concentration and grain size; *the ARM is a concentration-dependent parameter that is most sensitive to the smallest ferromagnetic particles* (Urbat et al., 2004).	Anhysteretic magnetizer (direct field ca 0.04 mT; a.c. field 100 mT) (Molspin Ltd)	$Am^2 kg^{-1}$
3.	χ_{FD} frequency dependent susceptibility	Susceptibility is measured in terms of frequency variations; frequency dependent susceptibility (χ_{FD}) was calculated as $\chi_{FD} = (\chi_{LF} - \chi_{HF})$.	Dual frequency susceptibility sensor (Bartington Instruments Ltd)	$\frac{\chi_{LF} - \chi_{HF}}{\chi_{LF}} \times 100 = \chi_{FD}$ % $m^3 kg^{-1}$
	IRM isothermal remanent magnetization	Reflects magnetic moment of plant leaves; each IRM value was determined by taking the mean of three acquisition–measurement cycles.	2G Enterprises 755R cryogenic magnetometer; Molspin spinner magnetometer; Princeton Vibrating Sample Magnetometer Model 3900; ASC Scientific IM-10 impulse magnetizer	$Am^2 kg^{-1}$; A for plant leaves

Continued

Table 4.2 List of different magnetic parameters, related instrumentation, and units—Cont'd

S. No.	Parameters	Explanation	Instruments used	Unit
4.	SIRM saturation isothermal remanent magnetization	It is actually the highest level of magnetic remanence that can be induced in a particular sample through application of high magnetic field; Acts as an indicator of volume concentration as well as grain size; *Besides the mineral and particle size-specific shapes of the acquisition curves, IRM measurements also provide a measure of the relative concentration of remanent magnetic particles (saturation IRM); SIRM is mainly influenced by the concentration of low coercivity, magnetite-type, minerals, and high-coercivity, hematite-type, minerals*	Pulse magnetizer (ca maximum 880 mT), fluxgate magnetometer (Bartington Instruments Ltd); ASC scientific pulse magnetizer (model IM10–30) linked with Molspin spinner magnetometer; Molspin pulse magnetizer; Princeton Vibrating Sample Magnetometer Model 3900	$Am^2 kg^{-1}$
5.	SIRM/χ	Helpful in differentiating dominant magnetic grain size; detects the nature, i.e., paramagnetic (low/zero value) or other	Ratio	Ratio Am^{-1}
	χ_{ARM}/SIRM	Commonly used as magnetic grain size indicators	Boyh parameters may be measured by Molspin Minispin magnetometer	Ratio
Demagnetization Parameters				
6.	$(B_0)_{cr}$	Estimated by applying one or more reversed magnetic fields to a previously saturated sample; the reverse field strength (mT) required to return a magnetized sample from its SIRM to zero is termed as coercivity of remanence $(B_0)_{cr}$;	Pulse magnetizer (ca maximum 880 mT), fluxgate magnetometer (Bartington Instruments Ltd)	mT

7.	IRM-χ/SIRM	Used to express the loss of magnetization at selected reversed fields; value usually lies between +1 and −1.	Pulse magnetizer (ca maximum 880mT), fluxgate magnetometer (Bartington Instruments Ltd)	Ratio
8.	"S"	Ratio obtained with IRM-$_{100mT}$ (a backfield differentiating ferromagnetic and antiferromagnetic mineral types)	Pulse magnetizer (ca maximum 880mT), fluxgate magnetometer (Bartington Instruments Ltd)	Ratio
9.	2D-magnetization	Calculated as the magnetic moment per leaf area	CCL cryogenic magnetometer	Amperes (A) ($A = Am^2/m^2$)

Modified after Rai (2013).

REFERENCES

Abbas, et al., 2009. Air pollution particulate matter ($PM_{2.5}$)-induced gene expression of volatile organic compound and/or polycyclic aromatic hydrocarbon-metabolizing enzymes in an in vitro coculture lung model. Toxicology in Vitro 23, 37–46.

Aboal, J.R., Fernandez, J.A., Carballeira, A., 2004. Oak leaves and pine needles as biomonitors of airborne trace elements pollution. Environmental and Experimental Botany 51, 215–225.

Adamowicz, V., Dales, R., Hale, B.,A., Hrudey, (Panel Chair), S.,E., Krupnick, A., Lippman, M., McConnell, J., Renzi, P., 2004. Report of an expert panel to review the socio-economic models and related components supporting the development of Canada-Wide Standards (CWS) for particulate matter (PM) and ozone to the Royal Society of Canada. Journal of Toxicology and Environmental Health, Part B 7 (3), 147–266.

Akbari, H., 2002. Shade trees reduce building energy use and CO_2 emissions from power plants. Environmental Pollution 116, S119–S126.

Akbari, H., Pomerantz, M., Taha, H., 2001. Cool surfaces and shade trees to reduce energy use and improve air quality in urban areas. Solar Energy 70, 295–310.

Al-Alawi, M., Batarseh, M.I., Carreras, H., Alawi, M., Jiries, A., Charlesworth, S.M., 2007. Aleppo pine bark as a biomonitor of atmospheric pollution in the arid environment of Jordan. Clean 35 (5), 438–443.

Al-Alawi, M.M., Mandiwana, K.L., 2007. The use of Aleppo pine needles as biomonitor of heavy metals in the atmosphere. Journal of Hazardous Materials 148, 43–46.

Alfani, A., Baldantoni, T., Maisto, G., Bartoli, G., Virzo De Santo, A., 2000. Temporal and spatial variation in C, N, S and trace element contents in the leaves of Quercus ilex within the urban area of Naples. Environmental Pollution 109, 119–129.

Alfani, A., Bartoli, G., Rutigliano, F.A., Maisto, G., Virzo de Santo, A., 1996. Trace metal biomonitoring in the soil and the leaves of Quercus ilex in the urban area of Naples. Biological Trace Element Research 51, 117–131.

Alfani, A., Maisto, G., Prati, M.V., Baldantoni, D., 2001. Leaves of Quercus ilex L. as biomonitors of PAHs in the air of Naples (Italy). Atmospheric Environment 35, 3553–3559.

Batarseh, M., Ziadat, A., Al-Alawi, M., Berdanier, B., Jiries, A., 2008. The use of cypress tree bark as an environmental indicator of heavy metals deposition in Fuheis City, Jordan. International Journal of Environment and Pollution 33 (2/3), 207–217.

Beckett, K.B., Freer-Smith, P.H., Taylor, G., 1998. Urban woodlands: their role in reducing the effects of particulate pollution. Environmental Pollution 99, 347–360.

Bennett, J.P., Wetmore, C.M., 1999. Changes in element contents of selected lichens over 11 years in northern Minnesota, USA. Environmental and Experimental Botany 41, 75–82.

Bergamaschi, L., Rizzio, E., Giaveri, G., Loppi, S., Gallorini, M., 2007. Comparison between the accumulation capacity of four lichen species transplanted to an urban site. Environmental Pollution 148, 468–476.

Blundell, A., Hannam, J.A., Dearing, J.A., Boyle, J.F., 2009. Detecting atmospheric pollution in surface soils using magnetic measurements: a reappraisal using an England and Wales database. Environmental Pollution 157 (10), 2878–2890.

Borrego, C., Monteiro, E., Ferreira, J., Miranda, A.I., 2009. Forecasting human exposure to atmospheric pollutants in Portugal – a modelling approach. Atmospheric Environment 43, 5796–5806.

Bussotti, F., Cenni, E., Cozzi, A., Ferretti, M., 1997. The impact of geothermal power plants on forest vegetation. A case study at Travale (Tuscany, Central Italy). Environmental Monitoring and Assessment 45, 181–194.

Calamari, D., Bacci, E., Focardi, S., Gaggi, C., Morosini, M., Vighi, M., 1991. Role of plant biomass in the global environmental partitioning of chlorinated hydrocarbons. Environmental Science and Technology 25, 1489–1495.

Çelik, A., Kartal, A.A., Akdo_gan, A., Kaska, Y., 2005. Determining the heavy metal pollution in Denizli (Turkey) by using *Robinio pseudoacacia* L. Environmental International 31, 105–112.

Costa, D.L., 2004. Issues that must be addressed for risk assessment of mixed exposures: the U.S. EPA experience with air quality. Journal of Toxicology and Environmental Health, Part A 67 (3), 195–207.

Coyle, D., Stieb, D., Burnett, R., DeCivita, P., Krewski, D., Chen, Y., Thun, M., 2003. Impact of particulate air pollution on quality-adjusted life expectancy in Canada. Journal of Toxicology and Environmental Health, Part A 66 (19), 1847–1864.

Craig, L., Brook, J.R., Chiotti, Q., Croes, B., Gower, S., Hedley, A., Krewski, D., Krupnick, A., Krzyzanowski, M., Moran, M.D., Pennell, W., Samet, J.M., Schneider, J., Shortreed, J., Williams, M., 2008. Air pollution and public health: a guidance document for risk managers. Journal of Toxicology and Environmental Health, Part A 71 (9–10), 588–698.

DaMatta, F.M., Grandis, A., Arenque, B.C., Buckeridge, M.S., 2010. Impacts of climate changes on crop physiology and food quality. Food Research International 43, 1814–1823.

Davidson, D.A., Wilkinson, A.C., Blais, J.M., 2003. Orographic cold-trapping of persistent organic pollutants by vegetation in mountains of western Canada. Environmental Science and Technology 37, 209–215.

Dimitriou, K., Kassomenos, P.A., Paschalidou, A.K., 2013. Assessing the relative risk of daily mortality associated with short term exposure to air pollutants in the European Union through air quality indices. Ecological Indicators 27, 108–115.

Dorevitch, S., Karandikar, A., Washington, G.F., Walton, G.P., Anderson, R., Nickels, L., 2008. Efficacy of an outdoor air pollution education program in a community at risk for asthma morbidity. Journal of Asthma 45 (9), 839–844.

El-Fadel, M., Massoud, M., 2000. Particulate matter in urban areas: health-based economic assessment. The Science of the Total Environment 257, 133–146.

Eriksson, G., Jensen, S., Kylin, H., Strachan, W., 1989. The pine needle as a monitor of atmospheric pollution. Nature 341, 42–44.

Fowler, D., Cape, J.N., Unsworth, M.H., 1989. Deposition of atmospheric pollutants on forests. Philosophical Transactions of the Royal Society of London 324, 247–265.

Franzaring, J., 1997. Temperature and concentration effects in biomonitoring of organic air pollutants. Environmental Monitoring and Assessment 46, 209–220.

Freer-Smith, P.H., El-Khatib, A.A., Taylor, G., 2004. Capture of particulate pollution by trees: a comparison of species typical of semi-arid areas (*Ficus nitida* and *Eucalyptus globulus*) with European and North-American species. Water, Air, and Soil Pollution 155, 173–187.

Giertych, M.J., De Temmerman, L.O., Rachwal, L., 1997. Distribution of elements along the length of Scots pine needles in a heavily polluted and control environment. Tree Physiology 17, 697–703.

Gratani, L., Varone, L., 2006. Carbon sequestration by *Quercus ilex* L. and *Quercus pubescens* Willd. and their contribution to decreasing air temperature in Rome. Urban Ecosystems 9, 27–37.

Gratani, L., Varone, L., 2007. Plant crown traits and carbon sequestration capability by *Platanus hybrida* Brot. in Rome. Landscape and Urban Planning 81, 282–286.

Gupta, P., et al., 2006. Satellite remote sensing of particulate matter and air quality assessment over global cities. Atmospheric Environment 40, 5880–5892.

Hansard, R., Maher, B.A., Kinnersley, R., 2011. Biomagnetic monitoring of industry derived particulate pollution. Environmental Pollution 159, 1673–1681.

Harrison, R.M., Yin, J., 2000. Particulate matter in the atmosphere: which particle properties are important for its effects on health? The Science of the Total Environment 249, 85–101.

Helmisaari, H.S., Derome, J., Fritze, H., Nieminen, T., Palmgren, K., Salemaa, M., Vanha-Majamaa, I., 1995. Copper in Scots pine forest around a heavy metal smelter in south-western Finland. Water, Air, & Soil Pollution 85, 1727–1732.

Janhall, S., 2015. Review on urban vegetation and particle air pollution – deposition and dispersion. Atmospheric Environment 105, 130–137.

Jensen, S., Eriksson, G., Kylin, H., Strachan, W., 1992. Atmospheric pollution by persistent organic compounds: monitoring with pine needles. Chemosphere 24, 229–245.

Kan, H., Chen, B., 2004. Particulate air pollution in urban areas of Shanghai, China: health based economic assessment. Science of the Total Environment 322, 71–79.

Krommer, V., Zechmeister, H.G., Roder, I., Scharf, S., Hanus-Illnar, A., 2007. Monitoring atmospheric pollutants in the biosphere reserve Wienerwald by a combined approach of biomonitoring methods and technical measurements. Chemosphere 67, 1956–1966.

Kylin, H., Söderkvist, K., Undemann, A., Franich, R., 2002. Seasonal variation of the terpene content, an overlooked factor in the determination of environmental pollutants in pine needles. Bulletin of Environmental Contamination and Toxicology 68, 155–160.

Larsen, R.S., Bell, J.N.B., James, P.W., Chimonides, P.J., Rumsey, F.J., Tremper, A., Purvis, O.W., 2007. Lichen and bryophyte distribution on oak in London in relation to air pollution and bark acidity. Environmental Pollution 146, 332–340.

Lau, O.W., Luk, S.F., 2001. Leaves of *Bauhinia blakeana* as indicators of atmospheric pollution in Hong Kong. Atmospheric Environment 35, 3113–3120.

Lehndorff, E., Schwark, L., 2009. Biomonitoring airborne parent and alkylated three-ring PAHs in the Greater Cologne Conurbation II: regional distribution patterns. Environmental Pollution 157 (5), 1706–1713.

Li, N., Sioutas, C., Cho, A., et al., 2003. Ultrafine particulate pollutants induce oxidative stress and mitochondrial damage. Environmental Health Perspectives 111, 455–460.

Loppi, S., Frati, L., Paoli, L., Bigagli, V., Rossetti, C., Bruscoli, C., Corsini, A., 2004. Biodiversity of epiphytic lichens and heavy metal contents of *Flavoparmelia caperata* thalli as indicators of temporal variations of air pollution in the town of Montecatini Terme (central Italy). Science of the Total Environment 326, 113–122.

Loppi, S., Ivanov, D., Boccardi, R., 2002. Biodiversity of epiphytic lichens and air pollution in the town of Siena (Central Italy). Environmental Pollution 116, 123–128.

Loppi, S., Nelli, L., Ancora, S., Bargagli, R., 1997. Passive monitoring of trace elements by means of tree leaves, epiphytic lichens and bark substrate. Environmental Monitoring and Assessment 45, 81–88.

Maher, B.A., Mooreb, C., Matzka, J., 2008. Spatial variation in vehicle-derived metal pollution identified by magnetic and elemental analysis of roadside tree leaves. Atmospheric Environment 42, 364–373.

Markert, B., 1993. Plants as Biomonitors. Indicators for Heavy Metals in the Terrestrial Environment. VCH Verlagsgesellschaft, Weinheim, 645 pp.

Martin, M.H., Coughtrey, P.J., 1982. Biological Monitoring of Heavy Metal Pollution: Land and Air. Applied Science Publishers, New York.

Matzka, J., Maher, B.A., 1999. Magnetic biomonitoring of roadside tree leaves: identification of spatial and temporal variations in vehicle-derived particulates. Atmospheric Environment 33, 4565–4569.

McClellan, R.O., 2002. Setting ambient air quality standards for particulate matter. Toxicology 181–182, 329–347.

McIntosh, G., Gómez-Paccard, M., Luisa Osete, M., 2007. The magnetic properties of particles deposited on *Platanus hispanica* leaves in Madrid, Spain, and their temporal and spatial variations. Science of the Total Environment 382, 135–146.

McPherson, E.G., Nowak, D., Heisler, G., Grimmond, S., Souch, C., Grant, R., Rowntree, R., 1997. Quantifying urban forest structure, function, and value: the Chicago urban forest climate project. Urban Ecosystems 1, 49–61.

McPherson, E.G., Scott, K.I., Simpson, J.R., 1998. Estimating cost effectiveness of residential yard trees for improving air quality in Sacramento, California, using existing models. Atmospheric Environment 32, 75–84.

McPherson, E.G., Simpson, J.R., Peper, P.J., Xiao, Q., 1999. Benefit-cost analysis of Modesto's municipal urban forest. Journal of Arboriculture 25, 235–248.

Mièieta, K., Murín, G., 1998. Three species of genus Pinus suitable as bioindicators of polluted environment. Water, Air, and Soil Pollution 104, 413–422.

Mingorance, M.D., Oliva, S.R., 2006. Heavy metal content in N. oleander leaves as urban pollution assessment. Environmental Monitoring and Assessment 119, 57–68.

Mitchell, G., Namdeo, A., Kay, D., 2000. A new disease-burden method for estimating the impact of outdoor air quality on human health. The Science of the Total Environment, 246, 153–163.

Mitchell, R., Maher, B., Kinnersley, R., 2010. Rates of particulate pollution deposition onto leaf surfaces: temporal and inter-species magnetic analyses. Environmental Pollution 158 (5), 1472–1478.

Mitchell, R., Maher, B.A., 2009. Evaluation and application of biomagnetic monitoring of traffic-derived particulate pollution. Atmospheric Environment 43, 2095–2103.

Monaci, F., Moni, F., Lanciotti, E., Grechi, D., Bargagli, R., 2000. Biomonitoring of airborne metals in urban environments: new tracers of vehicle emission, in place of lead. Environmental Pollution 107, 321–327.

Monteiro, A., Vautard, R., Lopes, M., Miranda, A.I., Borrego, C., 2005. Air pollution forecast in Portugal: a demand from the new air quality framework directive. International Journal of Environment and Pollution 25 (2), 4–15.

Moraes, R.M., Delitti, W.B.C., Moraes, J.A.P.V., 2003. Gas exchange, growth, and chemical parameters in a native Atlantic forest tree species in polluted areas of Cubatao, Brazil. Ecotoxicology and Environmental Safety 54, 339–345.

Moraes, R.M., Klumpp, A., Furlan, C.M., Klumpp, G., Domingos, M., Rinaldi, M.C.S., Modesto, I.F., 2002. Tropical fruit trees as bioindicators of industrial air pollution in southeast Brazil. Environment International 28, 367–374.

Moreno, E., Sagnotti, L., Dinare_s-Turell, J., Winkler, A., Cascella, A., 2003. Biomonitoring of traffic air pollution in Rome using magnetic properties of tree leaves. Atmospheric Environment 37, 2967–2977.

Murphy, K.J., Alpert, P., Cosentino, D., 1999. Local impacts of a rural coal-burning generating station on lichen abundance in a New England forest. Environmental Pollution 105, 349–354.

Nali, C., Balducci, E., Frati, L., Paoli, L., Loppi, S., Lorenzini, G., 2007. Integrated biomonitoring of air quality with plants and lichens: a case study on ambient ozone from central Italy. Chemosphere 67, 2169–2176.

National Expert Group on Transboundary Air Pollution, 2001. Transboundary acidification, eutrophication and ground level ozone in the UK.

Nieminen, T., Helmisaari, H.S., 1996. Nutrient translocation in the foliage of Pinus sylvestris L. growing along a heavy metal pollution gradient. Tree Physiology 16, 825–831.

Nowak, D.J., 1994. Air pollution removal by Chicago's urban forest. In: McPherson, E.G., Nowak, D.J., Rowntree, R.A. (Eds.), Chicago's Urban Forest Ecosystem: Results of the Chicago Urban Forest Climate Project. USDA, Forest Service, Gen. Tech. Rep. NE-186, pp. 63–81.

Nowak, D.J., Civerolo, J.C., Rao, S.T., Sistla, G., Luley, C.J., Crane, D.E., 2000. A modeling study of the impact of urban trees on ozone. Atmospheric Environment 34, 1601–1613.

Nowak, D.J., Crane, D.E., Stevens, J.C., 2006. Air pollution removal by urban trees and shrubs in the United States. Urban Forestry and Urban Greening 4, 115–123.

Nowak, D.J., Crane, D.E., Stevens, J.C., Ibarra, M., 2002. Brooklyn's Urban Forest. USDA, Forest Service, Gen. Tech. Rep. NE-290, 109 pp.

Nowak, D.J., McHale, P.J., Ibarra, M., Crane, D., Stecens, J., Luley, C., 1998. Modeling the effects of urban vegetation on air pollution. In: Gryning, S., Chaumerliac, N. (Eds.), Air Pollution Modeling and Its Application XII. Plenum, New York, pp. 399–407.

Ockenden, W.A., Steinnes, E., Parker, C., Jones, K.C., 1998. Observations on persistent organic pollutants in plants: implications for their use as passive air samplers and for POP cycling. Environmental Science and Technology 32, 2721–2726.

Okona-Mensah, K.B., Battershill, J., Boobis, A., Fielder, R., 2005. An approach to investigating the importance of high potency polycyclic aromatic hydrocarbons (PAHs) in the induction of lung cancer by air pollution. Food and Chemical Toxicology 43 (7), 1103–1116.

Pandey, S.K., Tripathi, B.D., Prajapati, S.K., Mishra, V.K., Upadhyay, A.R., Rai, P.K., Sharma, A.P., 2005. Magnetic properties of vehicle derived particulates and amelioration by $Ficus$ $infectoria$: a keystone species. Ambio a Journal on Human Environment 34 (8), 645–647.

Pearce, D., Crowards, T., 1996. Particulate matter and human health in the United Kingdom. Energy Policy 24 (7), 609–619.

Rai, P.K., 2011a. Dust deposition capacity of certain roadside plants in Aizawl, Mizoram: implications for environmental geomagnetic studies. In: Dwivedi, S.B., et al. (Ed.), Recent Advances in Civil Engineering, pp. 66–73.

Rai, P.K., 2011b. Biomonitoring of particulates through magnetic properties of roadside plant leaves. In: Tiwari, D. (Ed.), Advances in Environmental Chemistry. Excel India Publishers, New Delhi, pp. 34–37.

Rai, P.K., 2013. Environmental magnetic studies of particulates with special reference to biomagnetic monitoring using roadside plant leaves. Atmospheric Environment 72, 113–129.

Rai, P.K., 2015. Multifaceted health impacts of Particulate Matter (PM) and its management: an overview. Environmental Skeptics and Critics 4 (1), 1–26.

Randolph, J., 2004. Environmental Land Use Planning and Management. Island Press, Washington, DC, 664 pp.

Rautio, P., Huttunen, S., Lamppu, J., 1998a. Effects of sulphur and heavy metal deposition on foliar chemistry of Scots pines in Finnish Lapland and on the Kola Peninsula. Chemosphere 36 (4–5), 979–984.

Rautio, P., Huttunen, S., Lamppu, J., 1998b. Seasonal foliar chemistry of northern Scots pine under sulphur and heavy metal pollution. Chemosphere 37 (2), 271–278.

Rautio, P., Huttunen, S., Lamppu, J., 1998c. Element concentrations in Scots pine needles on radial transects across a subartic area. Water, Air, and Soil Pollution 102, 389–405.

Rosenfeld, A.H., Romm, J.J., Akbari, H., Pomerantz, M., 1998. Cool communities: strategies for heat islands mitigation and smog reduction. Energy and Buildings 28, 51–62.

Samet, J., Krewski, D., 2007. Health effects associated with exposure to ambient air pollution. Journal of Toxicology and Environmental Health, Part A 70 (3), 227–242.

Sandstrom, T., Cassee, F.R., Salonen, R., Dybing, E., 2005. Recent outcomes in European multicentre projects on ambient particulate air pollution. Toxicology and Applied Pharmacology 207, S261–S268.

Scott, K.I., McPherson, E.G., Simpson, J.R., 1998. Air pollutant uptake by Sacramento's urban forest. Journal of Arboriculture 24, 224–233.

Sgrigna, G., Sæbø, A., Gawronski, S., Popek, R., Calfapietra, C., 2015. Particulate matter deposition on $Quercus$ $ilex$ leaves in an industrial city of central Italy. Environmental Pollution 197 (2015), 187–194.

Simonich, S., Hites, R., 1994. Importance of vegetation in removing polycyclic hydrocarbons from the atmosphere. Nature 70, 49–51.

Singh, S.N., Verma, A., 2007. Phytoremediation of air pollutants, a review. In: Singh, S.N., Tripathi, R.D. (Eds.), Environmental Bioremediation Technology, 1. Springer, Berlin Heidelberg, pp. 293–314.

Stocker, R.A., 2000. Chapter13. Methodology for determining wildfire and prescribed fire air quality impacts on areas in the western United States. Journal of Sustainable Forestry 11 (1), 311–328.

Stone, J.B., Mednick, A.C., Holloway, T., Spak, S.N., 2007. Is compact growth good for air quality? Journal of the American Planning Association 73, 404–418.

Sultan, Z.M., 2007. Estimates of associated outdoor particulate matter health risk and costs reductions from alternative building, ventilation and filtration scenarios. Science of the Total Environment 377, 1–11.

Taha, H., 1996. Modeling the impacts of increased urban vegetation on the ozone air quality in the South Coast Air Basin. Atmospheric Environment 30, 3423–3430.

Taha, H., 1997. Modeling the impacts of large-scale albedo changes on ozone air quality in the South Coast Air Basin. Atmospheric Environment 31, 1667–1676.

Tchepel, O., Penedo, A., Gomes, M., 2007. Assessment of population exposure to air pollution by benzene. International Journal of Hygiene and Environmental Health 210, 407–410.

Tiwary, A., et al., 2007. An integrated tool to assess the role of new planting in PM_{10} capture and the human health benefits: a case study in London. Environmental Pollution 157, 2645–2653.

Tretiach, M., Adamo, P., Bargagli, R., Baruffo, L., Carletti, L., Crisafulli, P., Giordano, S., Modenesi, P., Orlando, S., Pittao, E., 2007. Lichen and moss bags as monitoring devices in urban areas. Part I: influence of exposure on sample vitality. Environmental Pollution 146, 380–391.

Ugwuanyi, J.U., Obi, F.C., 2002. A survey of health effects of air pollution on peasant farmers in Benue State, Nigeria. International Journal of Environmental Studies 59 (6), 665–677.

Ukonmaanaho, L., Starr, M., Hirvi, J.-P., Kokko, A., Lahermo, P., Mannio, J., Paukola, T., Ruoho-Airola, T., Tanskanen, H., 1998. Heavy metal concentrations in various aqueous and biotic media in Finnish Integrated Monitoring catchments. Boreal Environment Research 3, 235–249.

Urbat, M., Lehndorff, E., Schwark, L., 2004. Biomonitoring of air quality in the Cologne Conurbation using pine needles as a passive sampler – part 1: magnetic properties. Atmospheric Environment 38, 3781–3792.

Walden, J., 1999. Sample collection and preparation. In: Walden, J., Oldfield, F., Smith, J.P. (Eds.), Environmental Magnetism: A Practical Guide. Quaternary Research Association, Cambridge, England, pp. 26–34. Technical guide no. 6.

Wilson, J.G., Kingham, S., Pearce, J., Sturman, A.P., 2005. A review of intraurban variations in particulate air pollution: implications for epidemiological research. Atmospheric Environment 39, 6444–6462.

Wolterbeek, H.T., Verburg, T.G., 2004. Atmospheric metal deposition in a moss data correlation study with mortality and disease in the Netherlands. Science of the Total Environment 319 (1–3), 53–64.

Yang, S.N., Connell, D.W., Hawker, D.W., Kayal, S.I., 1991. Polycyclic aromatic hydrocarbons in air, soil and vegetation of an urban roadway. The Science of the Total Environment 102, 229–240.

Zhang, W., Yu, L., Lu, M., Hutchinson, M., Feng, H., 2007. Magnetic approach to normalizing heavy metal concentrations for particle size effects in intertidal sediments in the Yangtze Estuary, China. Environmental Pollution 147, 238–244.

CHAPTER FIVE

Biomagnetic Monitoring of Particulate Pollution through Plant Leaves: An Overview

5.1 INTRODUCTION

It is well known that air pollution represents a threat to both the environment and human health, and it is estimated that millions of tons of toxic pollutants are released into the air each year. In the recent Anthropocene era, the rapid pace of industrialization and urbanization has given birth to dust or particulate matter (PM) pollution, the impact of which may be correlated with urban planning as well as the topography of the particular region (Rai, 2011a,b, 2013). Besides social and economic problems, the development model of the so-called Third World has caused serious degradation of air quality particularly in relation to huge emissions of PM and hence posed challenges in the research fields of atmospheric science and technology. Environmental contamination and human exposure with respect to dust or PM pollution have dramatically increased during the past 10 years (Faiz et al., 2009; Rai, 2011a,b, 2013). Industrial emissions and vehicle-derived pollutants simultaneously release deleterious fine-grained particulates and magnetic particles into the atmosphere. These magnetic particles are derived from the presence of iron, often a mix of strongly magnetic (magnetite-like) and weakly magnetic (hematite-like) iron oxides.

The concept of environmental magnetism as a proxy for atmospheric pollution levels has been reported by several researchers based on analysis of soils and street or roof dust; however, only a small number of studies have emphasized the use of roadside plant leaves in monitoring the dust. In the current phase of science and technology, roads act as reservoirs of PM. Roads have a wide variety of primary, or direct, ecological effects as well as secondary, or indirect, ecological effects on the landscapes that they penetrate (Coffin, 2007). The particles of dust that deposit from the atmosphere and accumulate along roadsides are called road dust particles and originate

Biomagnetic Monitoring of Particulate Matter
ISBN 978-0-12-805135-1
http://dx.doi.org/10.1016/B978-0-12-805135-1.00005-6

from the interaction of solid, liquid, and gaseous metals (Akhter and Madany, 1993; Faiz et al., 2009). Since the roadside vegetation obviously comes into direct contact with particulates, irrespective of the sources, it would be beneficial to investigate them in terms of pollution science, particularly the role of plant leaves. Further, the implication of the intake of dust particles with high concentrations of heavy metals poses potentially deleterious effects on the health of human beings (Faiz et al., 2009). Moreover, apart from human health implications, there may be concomitant multifaceted impacts of dust particles or PM on global climate (Maher, 2009; Rai, 2011a,b).

It is now well established through research that urban PM may also contain magnetic particles (Hunt et al., 1984; Flanders, 1994; Morris et al., 1995; Matzka and Maher, 1999; Petrovsky and Ellwood, 1999; Maher et al., 2008; Rai, 2011a,b). These particles are derived from the presence of iron impurities in fuels, which form upon combustion of a nonvolatile residue, often a mix of strongly magnetic (magnetite-like) and weakly magnetic (hematite-like) iron oxides. Magnetite has been identified specifically as a combustion-derived component of vehicle exhaust materials (Abdul-Razzaq and Gautam, 2001; Maher et al., 2008). Apart from vehicular emissions, other natural sources (rock dust, street dust, sediments, etc.) may also contribute to magnetic minerals in the atmosphere (Maher et al., 2008; Maher, 2009; Rai, 2011a,b).

Magnetic minerals, particularly those derived from vehicular combustion, have a size range of 0.1–0.7 μm (Pandey et al., 2005; Maher, 2009). This grain size is particularly dangerous to humans because of its ability to be inhaled into the lungs (Pandey et al., 2005; Maher, 2009; Hansard et al., 2011; Rai, 2011a,b). Further, Matzka and Maher (1999) found that the grain size of magnetic particles from vehicle emissions is on the order of 0.3–3 μm, a size of particular potential hazard to health. Iron often occurs as an impurity in fossil fuels during industrial, domestic, or vehicle combustion, which ultimately forms a nonvolatile residue, often comprising glassy spherules of magnetic nature, with easily measurable magnetization levels (Matzka and Maher, 1999). Also, combustion-related particles in vehicles, via exhaust emissions and abrasion or corrosion of engine and vehicle body materials, can generate nonspherical magnetite particles (Pandey et al., 2005; Maher, 2009).

Nevertheless, there is still a paucity of focused research work in the multifaceted environmental dimensions of magnetic monitoring, particularly biomagnetic monitoring of particulate pollution with roadside plant leaves, which has the potential to become a new frontier in the field of atmospheric science and technology.

Several researchers opined that environmental magnetism studies act as a proxy for vehicle-derived pollutants through roadside plant leaves (Maher, 2009). Moreover, magnetic properties of PM may also act as a valuable tool in assessing the phenomenon of atmospheric climate change through study of the Chinese Loess Plateau. Dust or aerosols may act as indicators as well as agents of climate change, through radiative, cloud condensation and ocean biogeochemical effects (Watkins and Maher, 2003; Watkins et al., 2007; Maher, 2011). However, this chapter, in general, will address the issues pertaining to biomonitoring of particulates through magnetic properties of roadside plant leaves.

Dust pollution in the atmosphere, particularly of pollutant particles below $10\,\mu m$ (PM_{10}), is of current concern worldwide due to adverse health effects associated with their inhalation (Calderón-Garcidueñas et al., 2004; Morris et al., 1995; Oberdörster, 2000; Pope et al., 2004; Faiz et al., 2009). Moreover, PM in dust is thought to be the most harmful pollution component widely present in the environment, with no known level at which adverse human health effects actually occur (Bealey et al., 2007).

As discussed earlier in this chapter, PM also is comprised of magnetic particles, therefore, it is necessary both to characterize them and to investigate their sources. This chapter will include an overview of issues discussed in previous chapters with special emphasis on a detailed review of biomagnetic monitoring through plant leaves.

5.2 SOURCES OF PARTICULATE MATTER AND CHARACTERIZATION OF MAGNETIC PARTICLES

As mentioned earlier, sources of particulate pollution may be natural or anthropogenic in nature. Emission sources may include natural processes such as wildfires, volcano eruptions, and dust storms. The magnetic particles derived from multifaceted resources may be ferromagnetic, antiferromagnetic, and ferrimagnetic depending on the nature of spin acquired on the application of magnetic fields (Maher et al., 2009a).

Biogenic ferrimagnets are also reported to be present in the organisms like termites (Maher, 1998) and bacteria (Fassbinder et al., 1990). Manmade pollution encompasses combustion processes used for heating, power production, industry, and traffic vehicles (Hansard et al., 2011). Road traffic is considered to be one of the major sources of environmental pollution in urban areas, whereas other anthropogenic activities like power plants, metallurgy, mining, and dust originating from fragile rocks are of minor importance (Bucko et al., 2010, 2011). Although vehicles are the prime source of particulates (Maricq, 1999; Maher et al., 2008), other sources may also come in to play depending on the geography of particular landscapes (Rai, 2011b).

It has been shown that vehicle-derived pollutants simultaneously release deleterious fine-grained particulates and magnetic particles into the atmosphere (Pandey et al., 2005). Apart from vehicle-derived particulates, street dust also contains larger particles of PM posing little health risk (Simonich and Hites, 1995; Rautio et al., 1998; Veijalainen, 1998; Steinnes et al., 2000; Bargagli, 1998; Wolterbeek, 2002; Urbat et al., 2004). In a case study on geochemical and mineral magnetic characterization of urban sediment particulates, Manchester, UK, by Robertson et al. (2003), largely ferrimagnetic multidomain mineral magnetic composition of the particulates were recorded, indicating inputs of anthropogenic origin, primarily particulates derived from automobiles.

Industrial activity such as burning of fossil fuels also produces magnetically enhanced particulates in the environment (Blundell et al., 2009; Hansard et al., 2011). Xia et al. (2008) showed that the magnetic assemblage in the dustfall, mainly originating by coal burning, is dominated by pseudo-single domain (PSD) magnetite associated with maghemite and hematite. These particulates consist of coarse-grained multidomain and stable single domain magnetic minerals. The presence of magnetite as the dominant magnetic mineral has been confirmed by numerous analyses in different areas (Moreno et al., 2003; Urbat et al., 2004; Lehndorff et al., 2006; Maher, 2009; Saragnese et al., 2011; Hansard et al., 2011). In a case study on magnetic properties of roadside dust in Seoul, Korea, Kim et al. (2007) grouped magnetic materials into three types: (1) magnetic spherules possibly emitted from factories and domestic heating systems, (2) aggregates derived from vehicle emissions or motor vehicle brake systems, and (3) angular magnetic particles of natural origin. There may be several magnetic minerals associated with particulates (having different magnetic status) derived from terrestrial environment (Table 5.1).

Therefore, this chapter aims to describe the concept and multifaceted issues on magnetic properties of particulates as well as their environmental dimensions. Further, the chapter discusses the magnetic properties of plant leaves as a proxy for ambient particulate pollution. However, before going into details of the aforesaid, it is necessary to provide an overview of the health impacts of particulates, described in the next section. Here, it is worth mentioning that since the concept of biomagnetic monitoring has

Table 5.1 Description of several magnetic components (Rai, 2013)

Particulate fraction/ minerals associated	Formula	Magnetic status	Environmental associations
Magnetite	Fe_3O_4	Ferrimagnetic	Rare; derived from soil firing
Maghemite	γFe_2O_3	Ferrimagnetic	Usually occurs in extremely weathered tropical/ subtropical soils
Hematite	αFe_2O_3	Antiferromagnetic	Occurs in dry oxidized soils of relatively warmer regions
Goethite	$\alpha FeOOH$	Antiferromagnetic	Prevalent in moist soils of well-drained area
Ferrihydrite	$5Fe_2O_3 \cdot 9H_2O$	Paramagnetic	In poorly drained soils

After Schwertmann and Taylor (1977), Maher (1986), and Rai (2013).

proven to act as proxy for ambient PM, therefore, I mentioned the health implications of other fractions of PM in an extremely descriptive manner.

5.3 HUMAN HEALTH IMPACTS OF PARTICULATE POLLUTANTS

A growing body of literature has documented that particulate pollution causes adverse health impacts, particularly in the size range of less than 10 µm; this was briefly discussed in Chapter 2 (Curtis et al., 2006; Schwarze et al., 2006; Lipmann, 2007; Zeger et al., 2008; Mitchell et al., 2010; Rai, 2011a,b, 2013). It is now well established that particulate pollutants are associated with adverse effects on the respiratory system (Seaton et al., 1995; Schwartz, 1996; Pope et al., 2002; Knutsen et al., 2004; Knox, 2006; Maher et al., 2008; Hansard et al., 2011; Rai, 2011a,b, 2013). Further, particulates with aerodynamic diameter smaller than 2.5 mm ($PM_{2.5}$) have even more deleterious health impacts because when inhaled they penetrate deeper than PM_{10} and can reach the lung's alveoli (Rizzio et al., 1999; Harrison and Yin, 2000; Wichmann and Peters, 2000; Oberdörster et al., 2005; Moller et al., 2008; Saragnese et al., 2011). Links with lung cancer (Pope et al., 2002; Beeson et al., 1998) and increased cardiovascular mortality rates (Pope et al., 1995; Schwartz, 1996) have also been established. Children are particularly sensitive to air pollution as their lungs as well as immune systems are not completely developed when compared to adults (Bateson and Schwartz, 2007). Global records show that PM below size 2.5 µm causes 3%

of mortality from cardiopulmonary disease; 5% of mortality from cancer of the trachea, bronchus, and lung; and 10% of mortality from acute respiratory infections in children under age five (Cohen et al., 2005; Maher, 2009).

In the recent past, many studies highlighted the role of ambient airborne PM as an important environmental pollutant for many different cardiopulmonary diseases and lung cancers (Valavanidis et al., 2008). Further, it is increasingly being realized that generation of reactive oxygen species (ROS) and oxidative stress is an important toxicological mechanism of particle-induced lung cancer (Knaapen et al., 2004; Risom et al., 2005). The fraction of PM contains a number of constituents that may increase the generation of ROS by a variety of reactions such as transition metal catalyzes, metabolism, redox cycling of quinones, and inflammation. PM, thus, can generate oxidative damage to DNA, including guanine oxidation, which is mutagenic (Kasai, 1997; Moller et al., 2008).

In general, PM comprises polycyclic aromatic hydrocarbons (PAH) and volatile organic compounds (VOCs), which may have deleterious impact on human health (Lester and Seskin, 1970; Saldiva et al., 2002; Lin et al., 2003). The problem of metals is very much prevalent in the atmosphere particularly those in Asian countries, this being extensively reviewed by Fang et al. (2005). PM, especially traffic-related airborne particles, contains a large number of genotoxic/mutagenic chemical substances, which can cause DNA damage and promote malignant neoplasms (Ma et al., 1994; Grant, 1998; Johnson, 1998; Schoket, 1999; Lazutka et al., 1999; Sul et al., 2003; Claxton et al., 2004; Rajput and Agrawal, 2005; Claxton and Woodall, 2007; Lewtas, 2007; Valavanidis et al., 2008; Moller et al., 2008; Gammon and Santella, 2008; Coronas et al., 2009).

In view of the abovementioned deleterious impacts of PM, it is important to investigate the feasible and ecosustainable control technologies. The following is a discussion of a biomagnetic monitoring approach with roadside plant leaves.

5.4 GLOBAL RESEARCH ON ENVIRONMENTAL MAGNETISM

This section specifically discusses the magnetic properties of particulates and its environmental attributes with special reference to biomagnetic monitoring through plant leaves. We will discuss multidimensional attributes of environmental magnetism before focusing on biomagnetic monitoring.

With the advent of environmental magnetism, magnetic measurement is becoming an important means in particulate pollution study (Zhang et al., 2007). In environmental magnetism, there is growing interest in using magnetic methods in sediment tracing in the urban environment (Beckwith et al., 1990; Matzka and Maher, 1999; Xie et al., 2001).

In plants and soil samples, minerals capable of acquiring magnetic remanence include mainly the iron oxides (magnetite, maghemite, and hematite), oxyhydroxides (goethite), and sulfides (greigite). Magnetic iron sulfides are found only in reducing (anoxic) environments, such as estuarine muds, where organic matter is consumed by bacteria in the absence of oxygen. The strongest naturally occurring magnetic minerals are magnetite and maghemite, while hematite and goethite are magnetically much weaker (Maher, 2009).

The excellent potential of environmental magnetism as a proxy for atmospheric pollution levels has been reported by several researchers based on analysis of soils and street or roof dust (Hay et al., 1997; Hoffmann et al., 1999; Shu et al., 2000; Xie et al., 2000; Urbat et al., 2004), and vegetation samples including tree bark (Kletetschka et al., 2003) and leaves or needles (Matzka and Maher, 1999; Jordanova et al., 2003; Moreno et al., 2003). In urban particulates, a strong correlation has been observed between magnetic susceptibility as well as remanence and PM_{10} concentrations (e.g., Morris et al., 1995; Muxworthy et al., 2003; Sagnotti et al., 2006; Szönyi et al., 2008; Sagnotti et al., 2009; Hansard et al., 2011), as a proxy for particulate pollution concentrations (Hansard et al., 2011).

Active sampling, i.e., through air filters, has been used to discriminate particle sources and compare magnetic data with geochemical and meteorological data (Shu et al., 2001; Muxworthy et al., 2001, 2003; Spassov et al., 2004; McIntosh et al., 2007). Passive methods include the study of soils and street dust (Hay et al., 1997; Hoffman et al., 1999; Xie et al., 2000; Urbat et al., 2004; Shilton et al., 2005; McIntosh et al., 2007) and natural surfaces such as tree bark, tree leaves, and pine needles (e.g., Flanders, 1994; Matzka and Maher, 1999; Hanesch et al., 2003; Moreno et al., 2003; Urbat et al., 2004; McIntosh et al., 2007).

Blaha et al. (2008) analyzed fly ash samples from a black coal–fired power plant in Germany through the comparison of the bulk sample grain-size (0.5–300 μm) and grain-size spectra from magnetic extracts (1–186.5 μm) and showed that strongly magnetic particles mainly occur in the fine fractions of <63 μm.

Although environmental magnetism parameters have been optimized as qualitative proxy indicators of the distribution of anthropogenic particulates, heavy metals, and organic materials, Kim et al. (2009) proposed a quantitative magnetic proxy that is suitable for the monitoring of spatial and temporal pollution patterns in urban areas. In the aforesaid study, performed in southwestern Seoul, Kim et al. (2009) analyzed road dust samples with thermomagnetic data in conjunction with intensive electron microscopy and found predominance of carbon-bearing iron oxides, indicating that anthropogenic particulates mostly originated from fossil fuel combustions.

Muxworthy et al. (2003) advocated that saturation isothermal remanent magnetization (SIRM) was found to be strongly correlated with the PM mass, and not only acts as a proxy for PM monitoring but also is a viable alternative to magnetic susceptibility when the samples are magnetically too weak.

In several studies (Beckwith et al., 1986; Brilhante et al., 1989; Charlesworth and Lees, 1997; Xie et al., 2001), there are reports of possible links between magnetic properties and heavy metals in street dust. Moreover, in aerosols, magnetite is associated with heavy metals, e.g., zinc, cadmium, and chromium (Georgeaud et al., 1997) and mutagenic organic compounds (Morris et al., 1995), also dangerous to human health (Moreno et al., 2003). A significant correlation between sample mutagenicity and magnetic susceptibility for urban dust samples has already been established (Morris et al., 1995). Traditional geochemical methods (e.g., atomic absorption spectrophotometer (AAS), inductively coupled plasma spectroscopy (ICP-MS)) are relatively complex, time consuming, and expensive, and are therefore not suitable for performing mapping or monitoring of large-scale heavy metal or sediment pollution (Zhang et al., 2011).

Vehicle-derived pollutants are a major source of pollutants in the landscapes where intensive industries are not present. It has been shown that vehicle-derived pollutants simultaneously release deleterious fine-grained particulates and magnetic particles into the atmosphere (Pandey et al., 2005; Maher, 2009). Xie et al. (2001) in his investigation on Liverpool street dust suggested that magnetic properties, and mean values of some element concentrations and organic matter content, may be obtained with a small number of samples from a sampling period of one or several days.

Hoffmann et al. (1999) magnetically mapped soil surface emanating from vehicle pollution by measuring profiles of magnetic susceptibility along a German motorway. Another study on magnetic properties of dusts was done in the city of Munich, demonstrating a high correlation between total PM_{10} dust mass and its magnetic concentration as revealed by having high SIRM, the magnetization retained by a sample after exposure to a large magnetic field, e.g., 300 mT or 1T (Matzka, 1997; Matzka and Maher, 1999).

Industrial sources, e.g., thermal power plants, emit fly ashes, which also contribute to higher magnetic values (Schadlich et al., 1995; Pandey et al., 2009; Sharma and Tripathi, 2008; Hansard et al., 2011).

Environmental magnetic proxies provide a rapid means of assessing the degree of industrial heavy metal pollution in air, soils, and sediments (Zhang et al., 2011). Roadside dusts act as a common source for the heavy metals and magnetic carriers as revealed by a strong positive intercorrelation between the concentrations of heavy metals (Fe, Mn, Cr, Zn, Pb, and Cu) and magnetic susceptibility (Lu et al., 2008).

The association/correlation of magnetic properties with heavy metals may be demonstrated in soil samples (Lu et al., 2008). The magnetic parameters could provide a proxy measure for the level of heavy metal contamination and could be a potential tool for the detection and mapping of contaminated soils (Lu et al., 2008). Lu et al. (2008) investigated concentrations of copper and zinc and various magnetic parameters in contaminated urban roadside soils using chemical analysis and magnetic measurements, and their results revealed that high magnetic susceptibility may be attributed to anthropogenic soft ferrimagnetic particles. Hu et al. (2007) demonstrated significant correlations between heavy metals and several magnetic properties of the topsoil (from urban and agricultural sites) in Shanghai, indicating that the magnetic techniques can be used for monitoring soil pollution. Alagarsamy (2009) performed the environmental assessment of heavy metal concentrations and its impact in the coastal environment using magnetic techniques and found strong relationships between anhysteretic remanent magnetization (χARM) and heavy metals, which may be attributed to the role of iron oxides checking metal concentrations.

Road dust extracted from snow, collected near a busy urban highway and a low traffic road in a rural environment (southern Finland), was studied using magnetic, geochemical, and micromorphological analyses by Bucko et al. (2010). Results revealed a decreasing trend in χ and selected trace elements that was observed with increasing distance from the road edge.

Shilton et al. (2005) demonstrated significant correlations between the organic matter content of urban street dust and certain mineral magnetic properties, however, these researchers suggested that since the relationship may vary for different roads, even within same area, caution should be taken before concluding that magnetic parameters offer potential as a proxy for organic content.

McIntosh et al. (2007) found that concentration and grain-size trends across the roads act as the source of the magnetic signal, where the relationships between IRM1T (magnetic concentration) and the concentration of NOx and PM_{10} showed that the magnetic signal is specific to traffic-related emissions and not to total particle mass. Saragnese et al. (2011) investigated that superparamagnetic particles of nanometric dimension were identified in the PM by magnetic techniques and proposed a model linking total nitrogen oxides with magnetic particles.

In light of the previous discussion, it is quite clear that magnetic parameters may assist in multifaceted environmental geomagnetic studies (e.g., soil, street dust, sediments, etc.). The next part of chapter will mainly deal with biomagnetic monitoring through roadside plant leaves, which is the main focus of this chapter.

5.5 BIOMAGNETIC MONITORING OF PARTICULATES THROUGH ROADSIDE PLANT LEAVES

Several researchers have reported on the concept of environmental magnetism as a proxy for atmospheric pollution levels based on analysis of soils and street or roof dust (Hay et al., 1997; Hoffmann et al., 1999; Shu et al., 2000; Xie et al., 2000; Gautam et al., 2005; Hanesch et al., 2003; Jordanova et al., 2003; Urbat et al., 2004), and vegetation samples including tree bark (Kletetschka et al., 2003; Urbat et al., 2004). However, a cascade of research has emphasized the use of plant leaves in monitoring the dust (Matzka and Maher, 1999; Moreno et al., 2003; Jordanova et al., 2003; Urbat et al., 2004; Pandey et al., 2005; Maher et al., 2008; Maher, 2011; Rai, 2011b). Maher and her group were the leaders in performing numerous magnetic studies in relation to the environment, which eventually became the specialized discipline of environmental geomagnetism. Table 5.2 provides a list of plants used for biomagnetic monitoring of particulates. Biomagnetic monitoring studies were performed in mosses by a group of researchers (Vuković et al., 2015). Vuković et al. (2015) who demonstrated that moss bags can be effectively applied for biomagnetic monitoring of the spatiotemporal distribution of road traffic and vehicle-derived pollutants in urban areas.

It has been predicted that reductions in ambient particulate pollution concentrations of just $10\,\mu g\,m^{-3}$ are associated with increased life expectancy,

Table 5.2 Plants popularly used for biomagnetic monitoring

S. No.	Plants investigated	References
1.	*Populus deltoides* × *trichocarpa* "Beaupré"	Freer-Smith et al. (2005) and Rai (2013)
2.	*Ficus infectoria*	Pandey et al. (2005) and Rai (2013)
3.	*Dalbergia sissoo*	Prajapati et al. (2006) and Rai (2013)
4.	*Platanus* × *hispanica* leaves	McIntosh et al. (2007) and Rai (2013)
5.	*Platanus* sp.	Moreno et al. (2003) and Rai (2013)
6.	*Pinus nigra*/pine needles	Schadlich et al. (1995), Urbat et al. (2004), Lehndorff and Schwark (2004), Freer-Smith et al. (2005), Lehndorff et al. (2006), Lehndorff and Schwark, 2008, Lehndorff and Schwark (2009a,b), and Rai (2013)
7.	*Quercus ilex*	Moreno et al. (2003), Szönyi et al. (2008), and Rai (2013)
8.	*Quercus robur*	Mitchell et al. (2010) and Rai (2013)
9.	*Syzygium cumini*	Sharma et al. (2007) and Rai (2013)
10.	Birch (*Betula pendula*)	Matzka and Maher (1999), Hansard et al. (2011), Mitchell et al. (2010), and Rai (2013)
11.	Beech (*Fagus sylvatica*)	Hansard et al. (2011) and Rai (2013)
12.	Lime (*Tilia platyphyllos*)	Mitchell and Maher (2009), Mitchell et al. (2010), Maher et al. (2008), Hansard et al. (2011), and Rai (2013)
13.	Field maple (*Acer campestre*)	Freer-Smith et al. (2005), Hansard et al. (2011), and Rai (2013)
14.	Ash (*Fraxinus excelsior*)	Mitchell et al. (2010), Hansard et al. (2011), and Rai (2013)
15.	Sycamore (*Acer pseudoplatanus*)	Mitchell et al. (2010), Hansard et al. (2011), and Rai (2013)
16.	Elder (*Sambucus nigra*)	Mitchell et al. (2010), Hansard et al. (2011), and Rai (2013)
17.	Elm (*Ulmus procera*)	Mitchell et al. (2010), Hansard et al. (2011), and Rai (2013)
18.	Willow (*Salix alba*)	Mitchell et al. (2010), Hansard et al. (2011), and Rai (2013)
19.	*Salix matsudana*	Zhang et al. (2008) and Rai (2013)
20.	*Nerium* sp. (oleander)	Moreno et al. (2003) and Rai (2013)
21.	*Cercis siliquastrum*	Moreno et al. (2003) and Rai (2013)
22.	*Robinia pseudoacacia*	Moreno et al. (2003) and Rai (2013)
23.	Cypress (*Cupressus corneyana*)	Gautam et al. (2005) and Rai (2013)
24.	Silky oak (*Grevillea robusta*)	Gautam et al. (2005) and Rai (2013)
25.	Bottlebrush (*Callistemon lanceolatus*)	Gautam et al. (2005) and Rai (2013)
26.	Maple and *Acacia*	Jordanova et al. (2003) and Rai (2013)

independent of socioeconomic or demographic factors (Hansard et al., 2011). Henceforth, in view of the aforesaid, this section provides an overview of biomagnetic monitoring of roadside plant leaves as an integral part of the present chapter.

The fact that magnetic biomonitoring studies of plant leaves may act as a proxy of ambient particulate pollution is well proved now and has been emphasized in several places in this chapter. In conjunction with our brief discussion on the advantages of biomonitoring through magnetic properties in an earlier section, it is worth mentioning that magnetic biomonitoring of pollutants by measurements taken from roadside tree leaves is potentially efficient, as samples are abundant and hundreds of samples can be collected and analyzed within a few days' time (Rai, 2011b). Leaves with large surface areas per unit of weight, favorable surface properties (a waxy coating), and a long lifespan, such as conifer needles or evergreen tree leaves, are considered to be good accumulators of PM from the atmosphere (Freer-Smith et al., 1997; Alfani et al., 2000). Leaves are potentially efficient receptors and biomonitors of particulate pollution as they provide a large total surface for particle collection, numbers of samples and sample sites can be high (i.e., hundreds), and, in pollution contexts, the leaves themselves are insignificantly magnetic. Further, tree leaves also preclude sampling problems associated with the use of artificial particle collectors (including power requirements). Moreover, magnetic techniques are sensitive and rapid (e.g., Matzka and Maher, 1999; Muxworthy et al., 2003; Maher et al., 2008; Szönyi et al., 2008; Hansard et al., 2011). Magnetic measurements of leaves from several deciduous species can be intercalibrated (Mitchell et al., 2010), optimizing sampling density and resultant spatial resolution of the proxy PM_{10} data.

In view of these properties, magnetic biomonitoring studies of roadside plant leaves were performed in the Singrauli Industrial region (Pandey et al., 2005) and hilly areas of Nepal (Gautam et al., 2005), in addition to a series of pioneer works in European countries by a few group led by B.A. Maher (e.g., Matzka and Maher, 1999; Maher, 2009).

Magnetic properties of leaves (Muxworthy et al., 2002; Moreno et al., 2003; Urbat et al., 2004; Pandey et al., 2005; Maher et al., 2008; Mitchell and Maher, 2009; Maher, 2009) have been used to identify the spread of pollution derived from vehicular emissions. Therefore, the biomagnetic monitoring, using tree leaves as sampling surfaces, can generate high spatial-resolution PM_{10} proxy data (Hansard et al., 2011). Strong correlation has been demonstrated between magnetic properties, i.e., leaf

SIRM and/or magnetic susceptibility (χ) values and the amount of PM/ dust on the leaf surface (e.g., Halsall et al., 2008; Maher et al., 2008; Szönyi et al., 2008; Hansard et al., 2011). Correlations between magnetic parameters of plant leaves and toxic metals, such as lead, zinc, and iron, have also been investigated (e.g., Lu and Bai, 2006; Maher et al., 2008; Morton-Bermea et al., 2009; Hansard et al., 2011). Also, studies (Shu et al., 2001; Muxworthy et al., 2003; Saragnese et al., 2011) have found correlations between magnetic properties and levels of pollution (i.e., PM_{10} concentration and heavy metals).

While comparing the meteorological data and pollution data with reference to magnetic properties of urban particulates, Muxworthy et al. (2001) found that the magnetic hysteresis parameters generally had a stronger correlation with the meteorological data than with the pollution data.

Moreover, a biomagnetic monitoring approach may provide a robust means to achieve measurement and sourcing of PM_{10} at unprecedented levels of spatial resolution and is applicable all around the world (Maher, 2009) (e.g., Shu et al., 2001, in China; Gautam et al., 2005, in Nepal; Pandey et al., 2005, in India; Chaparro et al., 2006, in Argentina; Kim et al., 2007, in Korea; Szönyi et al., 2008, in Europe). Magnetic biomonitoring (Matzka and Maher, 1999; Maher, 2009) seems to be a valuable means both to gain significantly enhanced spatial resolution for pollutant data and to test proposed particulate source or health linkages.

In general, the magnetic properties of roadside tree leaves is greater when compared to those lying within the city center, as demonstrated in the case of birch leaves (Matzka and Maher, 1999). Several studies have investigated the biomonitoring of PAH (Lehndorff and Schwark, 2004; Lehndorff et al., 2006; Lehndorff and Schwark, 2009b) and trace elements/ heavy metals (Lehndorff and Schwark, 2008) in particulates through the study of magnetic properties in pine trees. A strong correlation between the magnetic susceptibility of pine needles and their metal (Fe) content has been demonstrated due to deposition of fly ash particles (Schadlich et al., 1995; Maher et al., 2008). A significant correlation was identified by Maher et al. (2008) between lead, iron, and leaf magnetic values in their study on PM.

Gautam et al. (2004) measured magnetic susceptibility of soils, sediments, and roadside materials, inside and outside the Kathmandu urban area, and magnetomineralogical analyses as well as scanning electron microscopy on magnetic extracts, grain-sized fractions or bulk samples of road dust and soils, suggesting lithogenic magnetite-like minerals and anthropogenic magnetic spherules to be the dominant contributors to the magnetic susceptibility signal. In addition to urban dust, research has indicated that lithogenic dust and incorporation of trace metals in the leaf tissue also control the magnetic susceptibility of tree leaves (Rodríguez-Germade et al., 2014). Climate, especially air humidity, and meteorology need to be considered when interpreting the magnetic properties of tree leaves as an atmospheric pollution tool (Rodríguez-Germade et al., 2014).

Hofman et al. (2013) investigated the usefulness of biomagnetic leaf monitoring of crown deposited particles to assess the spatial PM distribution inside individual tree crowns and an urban street canyon in Ghent (Belgium). Results demonstrated that biomagnetic monitoring can be used to assess spatial PM variations, even within single tree crowns (Hofman et al., 2013).

Zhang et al. (2008) used magnetic techniques including low-temperature experiments, successive acquisition of isothermal remanent magnetization (IRM), hysteresis loops, and measurements of saturated IRM (SIRM), which indicated that magnetic particles were omnipresent in tree bark and trunk wood of *Salix matsudana* and that these particles were predominantly magnetite with multidomain properties.

Hansard et al. (2011) investigated the utility of biomagnetic monitoring for spatial mapping of PM_{10} concentrations around a major industrial site and concluded the possibility of intercalibration by combining leaf magnetic measurements with colocated PM_{10} measurements.

5.6 RESEARCH STUDIES ON BIOMONITORING OF PARTICULATES THROUGH MAGNETIC PROPERTIES OF TREE LEAVES IN INDIA

In Indian cities airborne PM seems to be a very serious problem (Agarwal et al., 1999). Vegetation is an important tool to remove particulates from the atmosphere (Simonich and Hites, 1994). Several studies also revealed that leaves are sensitive and highly exposed parts of a plant and may act as persistent absorbers of dust in a polluted environment (Maiti, 1993; Samal and Santra, 2002). They act as pollution receptors and reduce dust concentration of the air (Maiti, 1993; Nowak, 1994; Singh, 2000; Samal and Santra, 2002). The capacity of leaves as dust receptors depends upon their surface geometry, phyllotaxy, epidermal and cuticular features, leaf pubescence, and the height and canopy of trees (Nowak, 1994; Singh, 2000; Singh et al., 2002). However, many plants are very sensitive to air pollutants, and pollutants can damage their leaves, impair plant growth, and limit primary productivity (Ulrich, 1984).

Thus, assessment of biochemical parameters is extremely relevant in order to identify tolerant plants. The most obvious damage occurs in the leaves and associated biochemical parameters. Growth and reproduction in some plants may be impaired and the populations of sensitive species are reduced while tolerant species can thrive and dominate the vegetation.

Likewise, a number of studies have been done on the pollution effects on different aspects of plant life such as overall growth and development (Gupta and Ghouse, 1987; Saquib et al., 1992; Misra and Behera, 1994; Pandey et al., 1999; Prusty et al., 2005), foliar morphology (Gupta and Mishra, 1994; Trivedi and Singh, 1995; Somashekar et al., 1999; Singh and Sthapak, 1999; Farooq et al., 2000; Pal et al., 2000; Shrivastava and Joshi, 2002), anatomy (Zafar, 1985; Arjunan et al., 1993; Garg et al., 2000; Singh, 2000), and biochemical changes (Balsberg-Pahlsson, 1989; Pandey and Sinha, 1991; Vyas et al., 1991; Budharaja and Agrawal, 1992; Tiwari and Patel, 1993; Krishnamurthy et al., 1994; Senapati and Misra, 1996; Pandey et al., 1999; Garty et al., 2001; Mashitha and Pise, 2001; Gavali et al., 2002). Pollution effects have been found to link the physiological response of plants to acceleration in the process of senescence (Lee et al., 1981; Kohert et al., 1986). One of the overt manifestations of plant senescence is gradual disappearance of chlorophyll and associated yellowing of leaves that may be linked with a consequent decline in the capacity for photosynthesis (Mandal and Mukherji, 2000).

This section of the chapter mentions some important case studies in India on magnetic properties of roadside plant leaves. These case studies were performed on selected roadside plants at a thermal power plant, a coal mine, and an urban area with heavy traffic.

In almost all of the case studies, the plant leaf samples were magnetized with a pulsed magnetic field of 300 mT by a Molspin pulse magnetizer (10–300 mT). The isothermal remanent magnetization (IRM300 mT) was then measured with a CCL cryogenic magnetometer having the sensitivity of 10^{-10} Am^2 (the weakest leaf samples had magnetizations of ~10^{-8} Am^2). The two-dimensional (2D) magnetization was calculated as the magnetic moment per leaf area, in units of amperes (A = Am^2/m^2). Therefore, all case studies adopted similar standard methodologies.

5.6.1 Research Studies in a Coal Mine and Thermal Power Plant of Singrauli Region, India

The Singrauli region (lies in 24°15′E to 82°40.9′N) lying between the states of Madhya Pradesh and Uttar Pradesh in northern India is now one of India's most important energy centers and is also called the "energy capital."

Eleven open-cast mining sites, occupying nearly 200 km^2, fuel six thermal power stations that generate 6800 MW or about 10% of India's installed generation capacity (Rai et al., 2007). However, the region is facing several environmental problems particularly in the context of pollution (Singh et al., 1991; Pandey et al., 2005; Rai, 2007a,b, 2008a,b,c,d,e, 2009, 2010a,b,c,d, 2012; Rai and Tripathi, 2007a,b, 2008, 2009; Rai et al., 2010).

1. Magnetic properties of *Ficus infectoria* in a coal mine area: Pakur. *F. infectoria*, an evergreen plant and a keystone species, was selected for magnetic properties in a coal mine area. IRM$_{300}$ and 2D-magnetization values were analyzed in a coal mine area of Singrauli region and are presented in Table 5.3 (Pandey et al., 2005). The four sampling sites were Gharsari village (remote rural area), Haul roads (near coal-handling plant), a road near a main substation, and Jawahar Colony (park). Vehicles using haul roads are mainly heavy-duty diesel vehicles that require more energy than private cars, thus increasing fuel consumption and exhaust emissions. Henceforth, 2D-magnetization values were recorded maximum adjacent to the haul road followed by the road near the main substation. Further, the study demonstrated that for leaves from individual trees, magnetization values fall from higher values on the road-proximal side of the tree to low values on the distal side, indicating the ability of trees to reduce particulate concentrations at respirable height within the atmosphere (Pandey et al., 2005). Sampling was carried out for five days to see the impact of meteorology. All five days show the same magnetization pattern, with higher values displayed by leaves from the trees adjacent to the haul road (Pandey et al., 2005).

Table 5.3 Biomagnetic monitoring study in Singrauli region A

Sampling site	2-D magnetization	2-D magnetization after cleaning of leaf (10^{-6} A)	% Magnetization removed by cleaning
1. Gharsari village (remote rural area)	15.64	4.85	69
2. Haul roads (near coal-handling plant)	84.52	10.14	88
3. Road near main substation	62.50	15	76
4. Jawahar Colony (park)	24.25	6.79	72

The different sampling days show not only the reproducibility of this magnetization pattern but also changes with time and weather. In the four sunny days, from day one to day two and day four to day five, the magnetization increased. After sampling on day two, there was rainfall in the area. Resampling on day three showed decline in the magnetization, suggesting the net removal of particles from leaf surfaces by rainwater (Pandey et al., 2005). Therefore, rainfall produces a net decrease in the concentrations of magnetic particles on leaf surfaces. Leaf drip and stem flow may thus act to remove fine-grained particulates from the roadside zone (Pandey et al., 2005).

2. Magnetic properties of *Syzygium cumini* (L.) in a thermal power plant area. The 2D magnetizations of leaves from *S. cumini* (L.) trees at different locations and distances from roads were performed by Sharma et al. (2007), and their background values, are shown in Table 5.4. The magnetization values were minimal for rural tree leaves (distant from major roads), higher for residential park areas (with less and light vehicular traffic), and greatest for tree leaves proximal to major highway roads (with high traffic of heavy diesel-driven vehicles) (Sharma et al., 2007). This magnetic pattern complies with the pattern reported by Pandey et al. (2005). Maximum 2D magnetizations were encountered for a sample taken from proximal to Anpara-Pipari National Highway and Anpara Market Road (with high diesel-driven generators and vehicular traffic load). For the distal, least magnetic samples, ~53% of the measured signal is derived from adhering

Table 5.4 Biomagnetic monitoring study in Singrauli region B

Sample location	2-D magnetization	2-D magnetization after cleaning of leaf (10^{-6} A)	% Magnetization removed by cleaning
1. Anpara colony (Park)	12.53	3.05	75.65
2. Anpara Market (Road)	63.18	12.75	79.81
3. Anpara Guest House (Park)	41.33	9.47	77.08
4. Proximal to Anpara-Pipari National Highway (near ATPS)	78.11	11.78	84.91
5. Distal to Anpara-Pipari National Highway (near ATPS)	54.96	9.43	82.84
6. Remote rural road (Dibulganj Village)	6.12	2.87	53.10

dust particles; for the proximal, most magnetic samples, ~85% is removed upon cleaning of the leaf (Table 5.4).

5.6.2 Case Studies in Urban Area of Varanasi, India

The city of Varanasi (82°15′E to 83°30′E and 24°35′N to 25°30′N, India), and its surrounding area, is characterized by little industry but, due to more than 1 million inhabitants, a substantial volume of traffic exists (Prajapati et al., 2006). The spatial and temporal variations of vehicle-derived particulates were mapped using magnetic analysis. Two-dimensional magnetization values were higher for leaves collected adjacent to major road sections than for those from village roads, suggesting vehicle emissions, rather than resuspended road dust, as the major cause of magnetic particles of roadside tree leaves. The magnetization of *Dalbergia sissoo* leaf samples collected from urban and rural areas is controlled by their proximity to major roads and may be an easily measurable proxy for vehicle-derived particulates. The highest leaf magnetizations were found adjacent to the major road of a national highway, indicating a combustion- and/or exhaust-related source of the magnetic particles (see Prajapati et al., 2006). The case study concluded that vehicle-derived particulates are responsible for tree leaf magnetism, and the leaf magnetizations values fall significantly from high values proximal to the roadside to lower values at the distal side (Prajapati et al., 2006).

5.7 IMPLICATIONS OF ENVIRONMENTAL GEOMAGNETISM/PALEOMAGNETISM IN CLIMATE CHANGE

As mentioned earlier in this chapter, atmospheric particulates or dust may play an active role in climate modification through radiative effects (Sokolik and Toon, 1999; Alfaro et al., 2004; IPCC, 2007; Maher, 2011) or indirectly via modification of clouds (Sassen et al., 2003; Spracklen et al., 2008; Maher, 2011) or of ocean uptake of atmospheric carbon dioxide (Martin, 1990; Bopp et al., 2003; Moore and Braucher, 2008; Maher, 2011). Unraveling the facts as well as factors related to climate change in the past provides the critical context and perspective for both the present understanding and future prediction of climate change (Maher, 2007).

Paleomagnetic studies particularly in the context of paleosols covering the phenomenon of soils and environment may provide an insight into past climate changes (Maher, 1984, 1986, 1998; Thompson and Oldfield, 1986; Fassbinder et al., 1990; Maher and Thompson, 1999; Maher et al., 1999,

2002, 2003; Maher and Hallam, 2005a,b; Alekseeva et al., 2007; Gibbard et al., 2010; Rai, 2011b). Thus, paleosols can act as integrative records of past climatic, lithological, geomorphological, geochemical, biological, and hydrological conditions (Wright, 1986; Retallack, 2001; Alekseeva et al., 2007). Quaternary aeolian sediments, worldwide in distribution, can comprise high-resolution archives of past climatic and environmental change, by incorporating chronological, physicochemical and magnetic information (Harrison et al., 2001; Kohfeld, 2002; Kohfeld et al., 2005; Maher, 2011).

Hematite and goethite have also been recognized as significant magnetic components of various terrestrial sediments (loess sequences of China) as well as marine ecosystems (Maher et al., 2004). It is generally assumed that increases in dust flux were globally "synchronous" and were on average two to five times greater than in interglacial stages (Kohfeld and Harrison, 2001; Maher, 2011). Loess sequences act both as dust sinks and as dust sources as demonstrated in by the Chinese Loess Plateau, which is a major dust source for the ocean sediments of the northwestern and equatorial Pacific (Rea and Hovan, 1995; Shigemitsu et al., 2007; Maher, 2011), the volcanic soils of Hawaii (Chadwick et al., 1999; Maher, 2011), and the accumulating ice caps of Greenland (Bory et al., 2003a,b; Maher, 2011). Henceforth, there exists a close linkage between magnetic properties of ultrafine soils/sediments/Chinese loess and climate change at regional or global levels (Maher, 1991, 2007, 2008, 2009, 2011; Maher and Thompson, 1992, 1995; Maher et al., 1994, 2003, 2004, 2009b, 2010; Rowe and Maher, 2000; Maher and Hu, 2006).

The magnetic properties are dominated by nonstoichiometric magnetite, with highest concentrations seen in PM_{10} (Revuelta et al., 2014). Low temperature magnetic analyses showed that the superparamagnetic fraction is more abundant when coarse, multidomain particles are present, confirming that they may occur as an oxidized outer shell around coarser grains (Revuelta et al., 2014).

A strong association of the magnetic parameters with a vehicular PM_{10} source has also been identified by Revuelta et al. (2014). Particulate matter can be defined by its aerodynamic properties. PM_{10} (the PM fraction with aerodynamic diameter <10 mm) is small enough to enter into the thoracic region, whereas $PM_{2.5}$ (<2.5 mm) and PM_1 (<1 mm) may penetrate further into the lungs and alveoli (Revuelta et al., 2014). A significant correlation demonstrated between the SIRM and weight of the surface-deposited particles confirms the potential of biomagnetic monitoring as a proxy for the amount of leaf-deposited particles (Hofman et al., 2014).

Maher et al. (2009b) investigated surface samples from the Gobi Desert, comparing their magnetic properties with those of the last glacial loess samples from across the Loess Plateau region, and demonstrated a magnetic mismatch between the Mongolian Gobi samples and the last glacial loess.

The magnetic technique appears to be useful for determining the anthropogenic pollution of Asian dust as demonstrated in a case study of South Korea (Kim et al., 2008). Magnetic properties of windblown dust and the Chinese Loess Plateau may be used to estimate rainfall for the last million years, thus assisting in study of the Southeast Asian summer monsoon (Maher and Hu, 2006; Maher, 2007). Maher (2007) mentioned that comparison of our magnetic rainfall record on land with environmental records from the deep sea shows that summer monsoon intensity is linked with growth and decay of continental-sized ice sheets, henceforth reflecting changes in the Earth's orbit around the Sun. The magnetic properties of the loess and paleosols of the Chinese Loess Plateau have been identified as a quantitative proxy of paleorainfall, through the Quaternary period (Heller et al., 1993; Maher et al., 1994; Liu et al., 1995; Han et al., 1996; Maher et al., 2003).

The magnetic properties of unpolluted soils appear to be causally related to rainfall and hence climate (Maher et al., 1994, 2003; Han et al., 1996). Further, these studies have demonstrated the presence of ultrafine-grained magnetite (inorganic and of in situ origin), which may have environmental implications particularly in the context of soil erosion and sedimentation (Maher and Taylor, 1988).

Magnetic properties appear to be a valuable tool to characterize and distinguish between the major suspended sediment inflows into an upland lake (Maher, 1988; Hatfield and Maher, 2008, 2009). In one case study on Bassenthwaite Lake in the United Kingdom, magnetic comparisons between the potential sources and the lake surface sediments indicated that Newlands Beck, providing only ~10% of the lake's hydraulic load, is the main contributor of sediment to the deep basin of the lake (Hatfield and Maher, 2008, 2009). Maher et al. (2009b) examined the utility of newly developed magnetic "fingerprinting" methods for identifying sediment provenance and its association with individual large flood events. Investigation of magnetic properties may reveal differences in both onshore and offshore sediment, thus, may have implications for source detection.

Further, the studies pertaining to magnetic susceptibility variation or magnetic properties in deep-sea sediments may be used as an indicator of climate change (Hounslow and Maher, 1999).

Magnetic/paleomagnetic studies also assist in tracing human evolution as demonstrated in the case of Luangwa Valley, Zambia, belonging to Eastern Africa. These studies as part of an integrated approach have broad implications for our understanding of early human behavior, adaptation, and survival, as well as the tempo and mode of colonization after humans' first dispersal out of Africa (Parfitt et al., 2010).

In a nutshell, the discussion in the present section reveals that magnetic studies have wider implications in the atmospheric environment.

5.8 CONCLUSIONS

Biomagnetic monitoring is very recent approach in the field of PM pollution science. With the advent of environmental magnetism, magnetic measurement is becoming an important means in particulate pollution study. Size fractionation of particulates reveal that aerodynamic diameters less than $10\,\mu m$ (PM_{10}) or $2.5\,mm$ ($PM_{2.5}$) have been reported to cause several health hazards particularly those of the respiratory tract. Further, PM may contain volatile organic compounds, polyaromatic hydrocarbons, and heavy metals that may cause genotoxic as well as mutagenic impacts. PM, irrespective of its source (natural or anthropogenic), is deleterious to human health. Currently, existing conventional technologies for monitoring of PMs are not totally feasible, which has paved the way for using the magnetic biomonitoring approach. Particulates deposited on plant leaves may be derived from iron impurities in fuels, industrial emissions, or rock dust, and magnetite or other magnetic spherules offer an opportunity for environmental geomagnetic studies (Figure 5.1). Environmental geomagnetic studies or specifically biomagnetic monitoring act as a proxy for vehicle- or industry-derived pollutants through roadside plant leaves on the one hand, while on the other hand, magnetic properties of several physical components of environment (dust, aerosols, Chinese Loess Plateau, soil, or marine sediments) may be linked to monitoring of climate change that occurred over a specified time period. The magnetic parameters of plant leaves (2D-magnetization values) may be correlated with the dust content and other constituents of dust (e.g., heavy metals). Thus, the investigation of magnetic properties of roadside plant leaves acts as an indicator or proxy for PM pollution, as demonstrated in several case studies mentioned in this chapter. However, the diversity of plants investigated for their biomagnetic monitoring potential is limited mostly to plants prevailing in temperate conditions, and therefore, the field is ripe for the

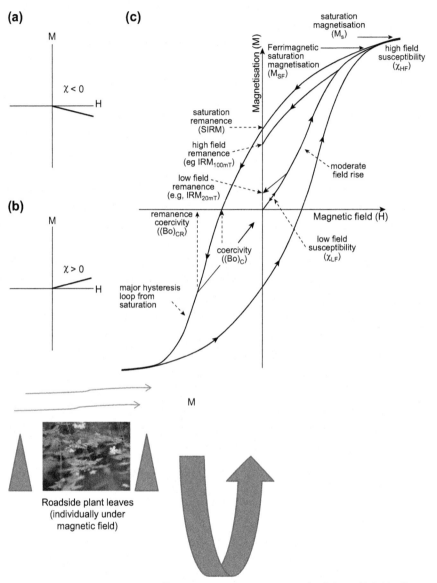

Figure 5.1 Demonstration of biomagnetic monitoring methodology (Rai, 2013).

investigations in the context of tropical plants. Janhall (2015) revealed that design and choice of urban vegetation is crucial when using vegetation as an ecosystem service for air quality improvements. The reduced mixing in trafficked street canyons on adding large trees increases local air pollution levels, while low vegetation close to sources can improve air quality by increasing deposition (Janhall, 2015).

Moreover, advancements in instrumentation or methodology related to magnetic studies may also improve the understanding of this research field, which is still in an embryonic stage. Henceforth, biomagnetic monitoring of particulate pollution may biomonitor the extent of dust/particulate pollution in an effective and systematic way. Nevertheless, there is still a paucity of focused research in the multifaceted environmental dimensions of magnetic monitoring, particularly biomagnetic monitoring of particulate pollution with roadside plant leaves. Hence, this area of interest possesses the potential to become a new frontier in the field of environmental science and technology.

REFERENCES

Abdul-Razzaq, W., Gautam, M., 2001. Discovery of magnetite in the exhausted material from a diesel engine. Applied Physics Letters 78, 2018–2019.

Agarwal, A., Narain, S., Sen, S., 1999. State of India's environment. In: Agarwal, A. (Ed.), The Citizens' Fifth Report. Part I: National Overview. Centre for Science and Environment, New Delhi, pp. 167–206.

Akhter, M.S., Madany, I.M., 1993. Heavy metals in street and house dust in Bahrain. Water, Air and Soil Pollution 66, 111–119.

Alagarsamy, R., 2009. Environmental magnetism and application in the continental shelf sediments of India. Marine Environmental Research 68 (2), 49–58.

Alekseeva, T., Alekseev, A., Maher, B.A., Demkin, V., 2007. Late Holocene climate reconstructions for the Russian steppe, based on mineralogical and magnetic properties of buried palaeosols. Palaeogeography Palaeoclimatology Palaeoecology 249, 103–127.

Alfani, A., Baldantoni, D., Maisto, G., Bartoli, G., Virzo de Santo, A., 2000. Temporal and spatial variation in C, N, S and trace element contents in the leaves of *Quercus ilex* within the urban area of Naples. Environmental Pollution 109, 119–129.

Alfaro, L., Chanda, A., Kalemli-Ozcan, S., Sayek, S., 2004. FDI and economic growth: the role of local financial markets. Journal of International Economics 64, 89–112.

Arjunan, M.C., Gunamani, T., Ponnammal, N.R., 1993. Pollution Research 12, 197–206.

Balsberg-Pahlsson, A.M., 1989. Effects of heavy-metal and SO$_2$ pollution on the concentrations of carbohydrates and nitrogen in tree leaves. Canadian Journal of Botany 67, 2106–2133.

Bargagli, R., 1998. Trace Elements in Terrestrial Plants: An Ecophysiological Approach to Biomonitoring and Biorecovery. Springer, Berlin. 324 pp.

Bateson, T.F., Schwartz, J., 2007. Children's response to air pollutants. Journal of Toxicology and Environmental Health Part A 71 (3), 238–243.

Bealey, W.J., Mcdonald, A.G., Nemitz, E., Donovan, R., Dragosits, U., Duffy, T.R., Fowler, D., 2007. Estimating the reduction of urban PM$_{10}$ concentrations by trees within an environmental information system for planners. Journal of Environmental Management 85, 44–58.

Beckwith, P.R., Ellis, J.B., Revitt, D.M., 1990. Applications of magnetic measurements to sediment tracing in urban highway environments. Science of the Total Environment 93, 449–463.

Beckwith, P.R., Ellis, J.B., Revitt, D.M., Oldfield, F., 1986. Heavy metal and magnetic relationships for urban source sediments. Physics of the Earth and Planetary Interiors 42, 67–75.

Beeson, W.L., Abbey, D.E., Knutsen, S.F., 1998. Long term concentrations of ambient air pollutants and incident lung cancer in Californian adults: results from the AHSMOG study. Environmental Health Perspectives 106, 813–822.

Blaha, U., Sapkota, B., Appel, E., Stanjek, H., Rösler, W., 2008. Micro-scale grain-size analysis and magnetic properties of coal-fired power plant fly ash and its relevance for environmental magnetic pollution studies. Atmospheric Environment 42 (36), 8359–8370.

Blundell, A., Hannam, J.A., Dearing, J.A., Boyle, J.F., 2009. Detecting atmospheric pollution in surface soils using magnetic measurements: a reappraisal using an England and Wales database. Environmental Pollution 157 (10), 2878–2890.

Bopp, L., Kohfeld, K.E., Le Quéré, C., Aumont, O., 2003. Dust impact on marine biota and atmospheric CO_2 during glacial periods. Paleoceanography 18. http://dx.doi.org/10.1029/2002PA000810. ISSN: 0883–8305.

Bory, A.J.M., Biscaye, P.E., Grousset, F.E., 2003a. Two distinct seasonal Asian source regions for mineral dust deposited in Greenland (NorthGRIP). Geophysical Research Letters 30.

Bory, A.J.M., Biscaye, P.E., Grousset, F.E., Zdanowicz, C.M., Prospero, J.M., 2003b. East-Asian dust sources and long-range transport: mineralogical and isotopic (Sr and Nd) constraints. Geochimica et Cosmochimica Acta 67, A43.

Brilhante, O., Daly, L., Trabuc, P., 1989. Application of magnetism to detect pollution caused by heavy metals in the environment. Comptes rendus de l'Académie des Sciences de Paris 309 (2), 2005–2012.

Bucko, M.S., Magiera, T., Johanson, Bo, Petrovský, E., Pesonen, L.J., 2011. Identification of magnetic particulates in road dust accumulated on roadside snow using magnetic, geochemical and micro-morphological analyses. Environmental Pollution 159, 1266–1276.

Bucko, M.S., Magiera, T., Pesonen, L.J., Janus, B., 2010. Magnetic, geochemical, and microstructural characteristics of road dust on roadsides with different traffic volumes-case study from Finland. Water Air and Soil Pollution 209, 295–306.

Budharaja, A., Agrawal, S.K., 1992. Acta Ecologia 14, 92–98.

Calderón-Garcidueñas, L., Reed, W., Maronpot, R.R., Henríquez-Roldán, C., Delgado-Chavez, R., Calderón-Garcidueñas, A., Dragustinovis, I., Franco-Lira, M., Aragón-Flores, M., Solt, A.C., Altenburg, M., Torres-Jardón, R., Swenburg, J.A., 2004. Brain inflammation and Alzheimer's-like pathology in individuals exposed to severe air pollution. Toxicologic Pathology 32, 650–658.

Chadwick, O.A., Derry, L.A., Vitousek, P.M., Huebert, B.J., Hedin, L.O., 1999. Changing sources of nutrients during four million years of ecosystem development. Nature 397, 491–497.

Chaparro, M.A.E., Gogorza, C.S.G., Chaparro, M.A.E., Irurzun, M.A., Sinito, A.M., 2006. Review of magnetism and heavy metal pollution studies of various environments in Argentina. Earth, Planets and Space 58, 1411–1422.

Charlesworth, S.M., Lees, J.A., 1997. The use of mineral magnetic measurements in polluted urban lakes and deposited dusts, Coventry, UK. Physics and Chemistry of the Earth 22, 203–206.

Claxton, L.D., Woodall Jr., G.M., 2007. A review of the mutagenicity and rodent carcinogenicity of ambient air. Mutation Research/Reviews in Mutation Research 636 (1–3), 36–94.

Claxton, L.D., Matthews, P.P., Warren, S.S., 2004. The genotoxicity of ambient outdoor air: a review: Salmonella mutagenicity. Mutation Research 567, 347–399.

Coffin, A.W., 2007. From roadkill to road ecology: a review of the ecological effects of roads. Journal of Transport Geography 15, 396–406.

Cohen, A.J., Anderson, H.R., Ostro, B., Pandey, K.D., Krzyzanowski, M., Kuenzli, N., 2005. The global burden of disease due to outdoor air pollution. Journal of Toxicology and Environmental Health, Part A: Current Issues 68 (13–14), 1301–1307.

Coronas, M.V., et al., 2009. Genetic biomonitoring of an urban population exposed to mutagenic airborne pollutants. Environment International 35 (7), 1023–1029.

Curtis, L., Rea, W., Smith-Willis, P., Fenyves, E., Pan, Y., 2006. Adverse health effects of outdoor pollutants. Environment International 32, 815–830.

Faiz, Y., Tufail, M., Javed, M.T., Chaudhry, M.M., Siddique, N., 2009. Road dust pollution of Cd, Cu, Ni, Pb and Zn along Islamabad Expressway, Pakistan. Microchemical Journal 92, 186–192.

Fang, G.C., Wu, Y.S., Huang, S.H., Rau, J.Y., 2005. Review of atmospheric metallic elements in Asia during 2000–2004. Atmospheric Environment 39 (17), 3003–3013.

Farooq, M., Arya, K.R., Kumar, S., Gopal, K., Joshi, P.C., Hans, R.K., 2000. Industrial pollutants mediated damage to mango (*Mangifera Indica*) crop: a case study. Journal of Environmental Biology 21, 165–167.

Fassbinder, J.W.E., Stanjek, H., Vali, H., 1990. Occurrence of magnetic bacteria in soil. Nature 343, 161–163.

Flanders, P.J., 1994. Collection, measurement, and analysis of airborne magnetic particulates from pollution in the environment. Journal of Applied Physics 75, 5931–5936.

Freer-Smith, P., Beckett, K., Taylor, G., 2005. Deposition velocities to *Sorbus aria, Acer campestre, Populus deltoides × trichocarpa* 'Beaupré', *Pinus nigra* and × *Cupressocyparis lelandii* for coarse, fine and ultra-fine particles in the urban environment. Environmental Pollution 133, 157–167.

Freer-Smith, P.H., Holloway, S., Goodman, A., 1997. The uptake of particulates by an urban woodland, site description and particulate composition. Environmental Pollution 95 (1), 27–35.

Gammon, M.D., Santella, R.M., 2008. PAH, genetic susceptibility and breast cancer risk: an update from the Long Island Breast Cancer Study Project. European Journal of Cancer 44 (5), 636–640.

Garg, S.S., Kumar, N., Das, G., 2000. Indian Journal of Environmental Protection 20, 326–328.

Garty, J., Tamir, O., Hassid, I., Eshel, A., Cohen, Y., Karnieli, A., Orlovsky, L., 2001. Photosynthesis, chlorophyll integrity, and spectral reflectance in lichens exposed to air pollution. Journal of Environmental Quality 30, 884–893.

Gautam, P., Blaha, U., Appel, E., 2005. Magnetic susceptibility of dust-loaded leaves as a proxy of traffic-related heavy metal pollution in Kathmandu city, Nepal. Atmospheric Environment 39, 2201–2211.

Gautam, P., Blaha, U., Appel, E., Neupane, G., 2004. Environmental magnetic approach towards the quantification of pollution in Kathmandu urban area, Nepal. Physics and Chemistry of the Earth, Parts A/B/C 29 (13–14), 973–984.

Gavali, J.G., Saha, D., Krishnayya, K., 2002. Difference in sulphur accumulation in eleven tropical tree species growing in polluted environments. Indian Journal of Environmental Health 44, 88–91.

Georgeaud, V.M., Rochette, P., Ambrosi, J.P., Vandamme, D., Williamson, D., 1997. Relationship between heavy metals and magnetic properties in a large polluted catchments, the Etang de Berre (South France). Physics and Chemistry of the Earth 22 (1–2), 211–214.

Gibbard, P.L., Boreham, S., Andrews, J.E., Maher, B.A., 2010. Sedimentation, geochemistry and palaeomagnetism of the West Runton Freshwater Bed, Norfolk, England. Quaternary International 228 (1–2), 8–20.

Grant, W.F., 1998. Higher plant assays for the detection of genotoxicity in air polluted environments. Ecosystem Health 4, 210–229.

Gupta, A.K., Mishra, R.M., 1994. Effect of lime kiln's air pollution on some plant species. Pollution Research 13, 1–9.

Gupta, M.C., Ghouse, A.K.M., 1987. The effect of coal smoke pollutants on growth yield and leaf epidermis features of *Abelmoschus esculenus* Moench. Environmental Pollution 43, 263–270.

Halsall, C.J., Maher, B.A., Karloukovski, V.V., Shah, P., Watkins, S.J., 2008. A novel approach to investigating indoor/outdoor pollution links: combined magnetic and PAH measurements. Atmospheric Environment 42, 8902–8909.

Han, J., Lu, H., Wu, N., Guo, Z., 1996. Magnetic susceptibility of modern soils in China and climate conditions. Studia Geophysica et Geodaetica 40, 262–275.

Hanesch, M., Scholger, R., Rey, D., 2003. Mapping dust distribution around an industrial site by measuring magnetic parameters of tree leaves. Atmospheric Environment 37, 5125–5133.

Hansard, R., Maher, B.A., Kinnersley, R., 2011. Biomagnetic monitoring of industry-derived particulate pollution. Environmental Pollution 159, 1673–1681.

Harrison, R.M., Yin, J.X., 2000. Particulate matter in the atmosphere: which particle properties are important for its effects on health? Science of the Total Environment 249, 85–101.

Harrison, S.P., Kohfeld, K.E., Roelandt, C., Claquin, T., 2001. The role of dust in climate changes today, at the last glacial maximum and in the future. Earth-Science Reviews 54, 43–80.

Hatfield, R.G., Maher, B.A., 2008. Suspended sediment characterization and tracing using a magnetic fingerprinting technique: Bassenthwaite Lake, Cumbria, UK. The Holocene 18 (1), 105–115.

Hatfield, R.G., Maher, B.A., 2009. Fingerprinting upland sediment sources: particle size-specific magnetic linkages between soils, lake sediments and suspended sediments. Earth Surface Processes and Landforms 34, 1359–1373.

Hay, K.L., Dearing, J.A., Baban, S.M.J., Loveland, P., 1997. A preliminary attempt to identify atmospherically derived pollution particles in English topsoils from magnetic susceptibility measurements. Physics and Chemistry of the Earth 22, 207–210.

Heller, F., Shen, C.D., Beer, J., Liu, X.M., Liu, T.S., Bronger, A., Suter, M., Bonani, G., 1993. Quantitative estimates of pedogenic ferromagnetic mineral formation in Chinese loess and palaeoclimatic implications. Earth and Planetary Science Letters 114, 385–390.

Hoffmann, V., Knab, M., Appel, E., 1999. Magnetic susceptibility mapping of roadside pollution. Journal of Geochemical Exploration 66, 313–326.

Hofman, J., Stokkaer, I., Snauwaert, L., Samson, R., 2013. Spatial distribution assessment of particulate matter in an urban street canyon using biomagnetic leaf monitoring of tree crown deposited particles. Environmental Pollution 183, 123–132.

Hofman, J., Wuyts, K., Wittenberghe, S.V., Brackx, M., Samson, R., 2014. On the link between biomagnetic monitoring and leaf-deposited dust load of urban trees: relationships and spatial variability of different particle size fractions. Environmental Pollution 189, 63–72.

Hounslow, M.W., Maher, B.A., 1999. Sources of the climate signal recorded by magnetic susceptibility variations in Indian Ocean sediments. Journal of Geophysical Research 104 (B3), 5047–5061.

Hu, X.F., Su, Y., Ye, R., Li, X.Q., Zhang, G.L., 2007. Magnetic properties of the urban soils in Shanghai and their environmental implications. CATENA 70 (3), 428–436.

Hunt, A., Jones, J., Oldfield, F., 1984. Magnetic measurements and heavy metals in atmospheric particulates of anthropogenic origin. Science of the Total Environment 33, 129–139.

IPCC, 2007. Climate Change 2007. Impacts, Adaptation and Vulnerability. Working Group II Contribution to the Fourth Assessment Report of the Intergovernmental Panel on Climate Change.

Janhall, S., 2015. Review on urban vegetation and particle air pollution – deposition and dispersion. Atmospheric Environment 105, 130–137.

Johnson, F.M., 1998. The genetic effects of environmental lead. Mutation Research/Reviews in Mutation Research 410 (2), 123–140.

Jordanova, N.V., Jordanova, D.V., Veneva, L., Yorova, K., Petrovsky, E., 2003. Magnetic response of soils and vegetation to heavy metal pollution—a case study. Environmental Science and Technology 37, 4417–4424.

Kasai, H., 1997. Analysis of a form of oxidative DNA damage, 8-hydroxy-2′-deoxyguanosine, as a marker of cellular oxidative stress during carcinogenesis. Mutation Research 387, 147–163.

Kim, W., Doh, S.-J., Park, Y.-H., Yun, S.-T., 2007. Two-year magnetic monitoring in conjunction with geochemical and electron microscopic data of roadside dust in Seoul, Korea. Atmospheric Environment 41, 7627–7641.

Kim, W., Doh, S.-J., Yu, Y., Lee, M., 2008. Role of Chinese wind-blown dust in enhancing environmental pollution in Metropolitan Seoul. Environmental Pollution 153, 333–341.

Kim, W., Doh, S.-J., Yu, Y., 2009. Anthropogenic contribution of magnetic particulates in urban roadside dust. Atmospheric Environment 43 (19), 3137–3144.

Kletetschka, G., Zila, V., Wasilewski, P.J., 2003. Magnetic anomalies on the tree trunks. Studia Geophysica et Geodaetica 47, 371–379.

Knaapen, A.M., Borm, P.J., Albrecht, C., Schins, R.P., 2004. Inhaled particles and lung cancer. Part A: mechanisms. International Journal of Cancer 109, 799–809.

Knox, E.G., 2006. Roads, railways and childhood cancers. Journal of Epidemiology and Community Health 60, 136–141.

Knutsen, S., Shavlik, D., Chen, L.H., Beeson, W.L., Ghamsary, M., Petersen, F., 2004. The association between ambient particulate air pollution levels and risk of cardiopulmonary and all-cause mortality during 22 years follow-up of a non-smoking cohort. Results from the AHSMOG study. Epidemiology 15, S45.

Kohert, R.J., Amundsun, R.G., Lawrence, J.A., 1986. Environmental Pollution, Series A 41, 219–234.

Kohfeld, K.E., 2002. The role of dust in climate cycles. Geochimica et Cosmochimica Acta 66, A409.

Kohfeld, K.E., Harrison, S.P., 2001. DIRTMAP: the geological record of dust. Earth-Science Reviews 54, 81–114.

Kohfeld, K.E., Le Quere, C., Harrison, S.P., Anderson, R.F., 2005. Role of marine biology in glacial–interglacial CO_2 cycles. Science 308, 74–78.

Krishnamurthy, R., Srinivas, T., Bhagwat, K.A., 1994. Journal of Environmental Biology 15, 97–106.

Lazutka, J.R., et al., 1999. Chromosomal aberrations and sister-chromatid exchanges in Lithuanian populations: effects of occupational and environmental exposures. Mutation Research/Genetic Toxicology and Environmental Mutagenesis 445 (2), 225–239.

Lee, E.H., Bennat, J.H., Heggestad, H.E., 1981. Retardation of senescence in red clover leaf discs by a new antiozonant, N-[2-(Oxo-1-imidazolidinyl) ethyl]-N-phenylurea. Plant Physiology 67, 347–350.

Lehndorff, E., Schwark, L., 2009a. Biomonitoring airborne parent and alkylated three-ring PAHs in the Greater Cologne Conurbation II: regional distribution patterns. Environmental Pollution 157 (5), 1706–1713.

Lehndorff, E., Schwark, L., 2009b. Biomonitoring airborne parent and alkylated three-ring PAHs in the Greater Cologne Conurbation I: temporal accumulation patterns. Environmental Pollution 157 (4), 1323–1331.

Lehndorff, E., Schwark, L., 2004. Biomonitoring of air quality in the Cologne conurbation using pine needles as a passive sampler: part II—PAH distribution. Atmospheric Environment 38 (23), 3793–3808.

Lehndorff, E., Schwark, L., 2008. Accumulation histories of major and trace elements on pine needles in the Cologne Conurbation as function of air quality. Atmospheric Environment 42 (5), 833–845.

Lehndorff, E., Urbat, M., Schwark, L., 2006. Accumulation histories of magnetic particles on pine needles as function of air quality. Atmospheric Environment 40 (36), 7082–7096.

Lester, B.L., Seskin, E.P., 1970. Air pollution and human health. American Association for the Advancement of Science 169, 723–733.

Lewtas, J., 2007. Air pollution combustion emissions: characterization of causative agents and mechanisms associated with cancer, reproductive, and cardiovascular effects. Mutation Research/Reviews in Mutation Research 636 (1–3), 95–133.

Lin, C.A., Pereira, L.A., Conceição, G.M.S., Kishi, H.S., Milani Jr., R., Braga, A.L.F., Saldiva, P.H.N., 2003. Association between air pollution and ischemic cardiovascular emergency room visits. Environmental Research 92, 57–63.

Lipmann, M., 2007. Health effects of airborne particulate matter. The New England Journal of Medicine 357 (23), 2395–2397.

Liu, X.M., Rolph, T., Bloemendal, J., Shaw, J., Liu, T.S., 1995. Quantitative estimates of palaeoprecipitation at Xifeng, in the Loess Plateau of China. Palaeogeography, Palaeoclimatology, Palaeoecology 113, 243–248.

Lu, J., Chen, G., Frierson, D.M.W., 2008. Response of the zonal mean atmospheric circulation to El Niño versus global warming. Journal of Climate 2, 5835–5851.

Lu, S.G., Bai, S.Q., 2006. Study on the correlation of magnetic properties and heavy metals content in urban soils of Hangzhou City, China. Journal of Applied Geophysics 60 (1), 1–12.

Ma, T.H., Cabrera, G.L., Chen, R., Gill, B.S., Sandhu, S.S., Valenberg, A.L., Salamone, M.F., 1994. Tradescantia micronucleus bioassay. Mutatation Research 310, 221–230.

Maher, B.A., 1984. Origins and Transformations of Magnetic Minerals in Soils (Ph.D. thesis). University of Liverpool.

Maher, B.A., 1991. Inorganic formation of ultrafine-grained magnetite. In: Frankel, R.B., Blakemore, R.P. (Eds.), Iron Biominerals. Plenum, New York, pp. 179–191.

Maher, B.A., Thompson, R., 1992. Paleoclimatic significance of the mineral magnetic record of the Chinese loess and paleosols. Quaternary Research 37, 155–170.

Maher, B.A., Thompson, R., 1995. Paleorainfall reconstructions from pedogenic magnetic susceptibility variations in the Chinese loess and paleosols. Quaternary Research 44, 383–391.

Maher, B.A., Thompson, R., Zhou, L.P., 1994. Spatial and temporal reconstructions of changes in the Asian palaeomonsoon: a new mineral magnetic approach. Earth Planetary Science letters 125, 461–471.

Maher, B.A., Taylor, R.M., 1988. Formation of ultrafine-grained magnetite in soils. Nature 336, 368–370.

Maher, B.A., Thompson, R., 1999. Palaeomonsoons I: the magnetic record of palaeoclimate in the terrestrial loess and palaeosol sequences. In: Maher, B.A., Thompson, R. (Eds.), Quaternary Climates, Environments and Magnetism. Cambridge University Press, Cambridge, pp. 81–125.

Maher, B.A., 1986. Characterisation of soils by mineral magnetic measurements. Physics of Earth and Planetary Interiors 42, 76–92.

Maher, B.A., 1988. Magnetic-properties of some synthetic sub-micron magnetites. Geophysical Journal-Oxford 94, 83–96.

Maher, B.A., 1998. Magnetic properties of modern soils and quaternary loessic paleosols: paleoclimatic implications. Palaeogeography, Palaeoclimatology, Palaeoecology 137, 25–54.

Maher, B.A., 2007. Environmental magnetism and climate change. Contemporary Physics 48, 247–274.

Maher, B.A., 2008. Holocene variability of the East Asian summer monsoon from Chinese cave records: a re-assessment. Holocene 18, 861–866.

Maher, B.A., 2009. Rain and dust: magnetic records of climate and pollution. Elements 5, 229–234.

Maher, B.A., 2011. The magnetic properties of quaternary aeolian dusts and sediments, and their palaeoclimatic significance. Aeolian Research. http://dx.doi.org/10.1016/j.aeolia.2011.01.005.

Maher, B.A., Alekseev, A., Alekseeva, T., 2002. Variation of soil magnetism across the Russian steppe: its significance for use of soil magnetism as a paleorainfall proxy. Quaternary Science Reviews 21, 1571–1576.

Maher, B.A., Hallam, D.F., 2005a. Palaeomagnetic correlation and dating of Plio/Pleistocene sediments at the southern margins of the North Sea basin: paper I. Journal of Quaternary Science 20, 67–77.

Maher, B.A., Hallam, D.F., 2005b. Magnetic carriers and remanence mechanisms in magnetite-poor sediments of Pleistocene age, southern North Sea margin: paper II. Journal of Quaternary Science 20, 79–94.

Maher, B.A., Hu, M.Y., 2006. A high-resolution record of Holocene rainfall variations from the western Chinese Loess Plateau: antiphase behaviour of the African/Indian and East Asian summer monsoons. Holocene 16, 309–319.

Maher, B.A., Hu, M.Y., Roberts, H.M., Wintle, A.G., 2003. Holocene loess accumulation and soil development at the western edge of the Chinese Loess Plateau: implications for magnetic proxies of palaeorainfall. Quaternary Science Reviews 22, 445–451.

Maher, B.A., Karloukovski, V.V., Mutch, T.J., 2004. High-field remanence properties of synthetic and natural submicrometre hematites and goethites: significance for environmental contexts. Earth and Planetary Science Letters 226, 491–505.

Maher, B.A., Mooreb, C., Matzka, J., 2008. Spatial variation in vehicle-derived metal pollution identified by magnetic and elemental analysis of roadside tree leaves. Atmospheric Environment 42, 364–373.

Maher, B.A., Watkins, S.J., Brunskill, G., Alexander, J., Fielding, C.R., 2009a. Sediment provenance in a tropical fluvial and marine context by magnetic 'fingerprinting' of transportable sand fractions. Sedimentology 56, 841–861.

Maher, B.A., Mutch, T.J., Cunningham, D., 2009b. Magnetic and geochemical characteristics of Gobi Desert surface sediments: implications for provenance of the Chinese Loess Plateau. Geology 37, 279–282.

Maher, B.A., Prospero, J.M., Mackie, D., Gaiero, D., Hesse, P.P., Balkanski, Y., 2010. Global connections between aeolian dust, climate and ocean biogeochemistry at the present day and at the last glacial maximum. Earth-Science Reviews 99, 61–97.

Maher, B.A., Thompson, R., Hounslow, M.W., 1999. Introduction to quaternary climates, environments and magnetism. In: Maher, B.A., Thompson, R. (Eds.), Quaternary Climates, Environments and Magnetism. Cambridge University Press, pp. 1–48.

Maiti, S.K., 1993. Indian Journal of Environmental Protection 13, 276–280.

Mandal, M., Mukherji, S., 2000. Changes in chlorophyll context, chlorophyllase activity, Hill reaction, photosynthetic CO_2 uptake, sugar and starch contents in five dicotyledonous plants exposed to automobile exhaust pollution. Journal of Environmental Biology 21, 37–41.

Maricq, M., 1999. Examination of the size-resolved and transient nature of motor vehicle particle emissions. Environmental Science and Technology 33, 1618–1626.

Martin, J.H., 1990. Glacial-interglacial CO_2 change: the iron hypothesis. Paleoceanography 5, 1–13.

Mashitha, P.M., Pise, V.I., 2001. Biomonitoring of air pollution by correcting the pollution tolerance index of some commonly ground trees of an urban area. Pollution Research 20, 195–197.

Matzka, J., 1997. Magnetische, elektronenmikroskopische und lichtmikroskopische Unter-suchungen an StaKuben und Aschen sowie an einzelen Aschepartikeln Diploma Thesis. Inst. Allg. Angew. Geophysik. University of MuK nchen, MuK nchen, Germany.

Matzka, J., Maher, B.A., 1999. Magnetic biomonitoring of roadside tree leaves: identification of spatial and temporal variations in vehicle-derived particulates. Atmospheric Environment 33, 4565–4569.

McIntosh, G., Gómez-Paccard, M., Luisa Osete, M., 2007. The magnetic properties of particles deposited on Platanus × hispanica leaves in Madrid, Spain, and their temporal and spatial variations. Science of the Total Environment 382, 135–146.

Misra, R., Behera, P.K., 1994. Pollution Research 13, 203–206.

Mitchell, R., Maher, B.A., 2009. Evaluation and application of biomagnetic monitoring of traffic-derived particulate pollution. Atmospheric Environment 43, 2095–2103.

Mitchell, R., Maher, B., Kinnersley, R., 2010. Rates of particulate pollution deposition onto leaf surfaces: temporal and inter-species magnetic analyses. Environmental Pollution 158 (5), 1472–1478.

Moller, P., et al., 2008. Air pollution, oxidative damage to DNA, and carcinogenesis. Cancer Letters 266, 84–97.

Moore, J.K., Braucher, O., 2008. Sedimentary and mineral dust sources of dissolved iron to the world ocean. Biogeosciences 5, 631–656.

Moreno, E., Sagnotti, L., Dinarès-Turell, J., Winkler, A., Cascella, A., 2003. Biomonitoring of traffic air pollution in Rome using magnetic properties of tree leaves. Atmospheric Environment 37, 2967–2977.

Morris, W.A., Versteeg, J.K., Bryant, D.W., Legzdins, A.E., Mccarry, B.E., Marvin, C.H., 1995. Preliminary comparisons between mutagenicity and magnetic susceptibility of respirable airborne particulate. Atmospheric Environment 29, 3441–3450.

Morton-Bermea, O., Hernandez, E., Martinez-Pichardo, E., Soler-Arechalde, A.M., Lozano Santa-Cruz, R., Gonzalez-Hernandez, G., Beramendi-Orosco, L., Urrutia- Fucugauchi, J., 2009. Mexico City topsoils: heavy metals vs. magnetic susceptibility. Geoderma 151 (3–4), 121–125.

Muxworthy, A.R., Matzka, J., Peterson, N., 2001. Comparison of magnetic parameters of urban particulate matter with pollution and meteorological data. Atmospheric Environment 35, 4379–4386.

Muxworthy, A.R., Matzka, J., Davila, A.F., Petersen, N., 2003. Magnetic signature of daily sampled urban atmospheric particles. Atmospheric Environment 37, 4163–4169.

Muxworthy, A.R., Schmidbauer, E., Petersen, N., 2002. Magnetic properties and Mössbauer spectra of urban atmospheric particulate matter: a case study from Munich, Germany. Geophysical Journal International 150, 558–570.

Nowak, D.J., 1994. Air pollution removal by Chicago's urban forest. In: McPherson, E.G., Nowak, D.J., Rowntree, R.A. (Eds.), Chicago's Urban Forest Ecosystem: Results of the Chicago Urban Forest Climate Project. USDA, Forest Service, Gen. Tech. Rep. NE-186, pp. 63–81.

Oberdörster, G., 2000. Toxicology of ultrafine particles: in vivo studies. Philosophical Transactions of the Royal Society of London, A 358, 2719–2740.

Oberdörster, G., Oberörster, E., Oberdörster, J., 2005. Nanotoxicology: an emerging discipline evolving from studies of ultrafine particles. Environmental Health Perspective 113, 823–839.

Pal, A., Kulshreshtha, K., Ahmad, K.J., Yunus, M., 2000. Changes in leaf surface structures of two avenue tree species caused by auto exhaust pollution. Journal of Environmental Biology 21, 15–21.

Pandey, D.D., Sinha, C.S., 1991. Environment Ecology 9, 617–620.

Pandey, D.D., Nirala, K., Gautam, R.R., 1999. Indian Journal of Environment and Ecoplanning 2, 43–46.

Pandey, J., Pandey, R., Shubhashish, K., 2009. Air-borne heavy metal contamination to dietary vegetables: a case study from India. Bulletin of Environmental Contamination and Toxicology 83, 931–936.

Pandey, S.K., Tripathi, B.D., Prajapati, S.K., Mishra, V.K., Upadhyay, A.R., Rai, P.K., Sharma, A.P., 2005. Magnetic properties of vehicle derived particulates and amelioration by *Ficus infectoria*: a keystone species. AMBIO: A Journal on Human Environment 34 (8), 645–647.

Parfitt, et al., 2010. Early Pleistocene human occupation at the edge of the boreal zone in northwest Europe. Nature 466, 229–233.

Petrovsky, E., Ellwood, B.B., 1999. Magnetic monitoring of pollution of air, land and waters. In: Maher, B.A., Thompson, R. (Eds.), Quaternary Climates, Environments and Magnetism. Cambridge University Press, Cambridge, pp. 279–322.

Pope III, C.A., Burnett, R.T., Thun, M.J., Calle, E.E., Krewski, D., Kazuhiko, I., Thurston, G.D., 2002. Lung cancer, cardiopulmonary mortality, and long-term exposure to fine particulate air pollution. Journal of American Medical Association 287, 1132–1141.

Pope, C.A., Hansen, M.L., Long, R.W., Nielsen, K.R., Eatough, N.L., Wilson, W.E., Eatough, D.J., 2004. Ambient particulate air pollution, heart rate variability, and blood markers of inflammation in a panel of elderly subjects. Environmental Health Perspectives 112, 339–345.

Pope, C.A., Thun, M., Namboodiri, M., Dockery, D., Evans, J., Speizer, F., Heath, C., 1995. Particulate air pollution as a predictor of mortality in a prospective study of U.S. adults. American Journal Respiratory Critical Care Medicine 151, 669–674.

Prajapati, S.K., Pandey, S.K., Tripathi, B.D., 2006. Magnetic biomonitoring of roadside tree leaves as a proxy of vehicular pollution. Environmental Monitoring and Assessment 120, 169–175.

Prusty, B.A.K., Mishra, P.C., Azeezb, P.A., 2005. Dust accumulation and leaf pigment content in vegetation near the national highway at Sambalpur, Orissa, India. Ecotoxicology and Environmental Safety 60, 228–235.

Rai, P.K., 2010a. Microcosm investigation on phytoremediation of Cr using *Azolla pinnata*. International Journal of Phytoremediation 12, 96–104.

Rai, P.K., 2012. An eco-sustainable green approach for heavy metals management: two case studies of developing industrial region. Environmental Monitoring and Assessment. http://dx.doi.org/10.1007/s10661-011-1978-x Published online 5 April, 2011.

Rai, P.K., Tripathi, B.D., 2007a. Heavy metals removal using nuisance blue green alga *Microcystis* in continuous culture experiment. Environmental Sciences 4 (1), 53–59.

Rai, P.K., Tripathi, B.D., 2007b. Microbial contamination in vegetables due to irrigation with partially treated municipal wastewater in a tropical city. International Journal of Environmental Health Research 17 (5), 389–395.

Rai, P.K., Tripathi, B.D., 2008. Heavy metals in industrial wastewater, soil and vegetables in Lohta village, India. Toxicological and Environmental Chemistry 90 (2), 247–257.

Rai, P.K., Tripathi, B.D., 2009. Comparative assessment of *Azolla pinnata* and *Vallisneria spiralis* in Hg removal from G.B. Pant Sagar of Singrauli industrial region, India. Environmental Monitoring and Assessment 148, 75–84.

Rai, P.K., 2007a. Phytoremediation of Pb and Ni from industrial effluents using *Lemna minor*: an eco-sustainable approach. Bulletin of Bioscience 5 (1), 67–73.

Rai, P.K., 2007b. Wastewater management through biomass of *Azolla pinnata*: an eco-sustainable approach. AMBIO 36 (5), 426–428.

Rai, P.K., 2008a. Heavy-metal pollution in aquatic ecosystems and its phytoremediation using wetland plants: an eco-sustainable approach. International Journal of Phytoremediation 10 (2), 133–160.

Rai, P.K., 2008b. Mercury pollution from chlor-alkali industry in a tropical lake and its biomagnification in aquatic biota: link between chemical pollution, biomarkers and human health concern. Human and Ecological Risk Assessment: An International Journal 14, 1318–1329.

Rai, P.K., 2008c. Phytoremediation of Hg and Cd from industrial effluents using an aquatic free floating macrophyte *Azolla pinnata*. International Journal of Phytoremediation 10 (5), 430–439.

Rai, P.K., 2008d. Heavy metals in water, sediments and wetland plants in an aquatic ecosystem of tropical industrial region, India. Last update 2009 Environment Monitoring and Assessment 158, 433–457. http://dx.doi.org/10.1007/s10661-008-0595-9.

Rai, P.K., 2008e. Ecological Investigation on Heavy Metal Pollution of G.B. Pant Sagar and its Phytoremediation (Ph.D. dissertation). Banaras Hindu University, Varanasi, India.

Rai, P.K., 2009. Heavy metal phytoremediation from aquatic ecosystems with special reference to macrophytes. Critical Review in Environmental Science and Technology 39 (9), 697–753.

Rai, P.K., 2010b. Phytoremediation of heavy metals in a tropical impoundment of industrial region. Environment Monitoring and Assessment 165, 529–537.

Rai, P.K., 2010c. Seasonal monitoring of heavy metals and physico-chemical characteristics in a lentic ecosystem of sub-tropical industrial region, India. Environmental Monitoring and Assessment 165, 407–433.

Rai, P.K., 2010d. Heavy metal pollution in lentic ecosystem of sub-tropical industrial region and its phytoremediation. International Journal of Phytoremediation 12 (3), 226–242.

Rai, P.K., 2011a. Dust deposition capacity of certain roadside plants in Aizawl, Mizoram: Implications for environmental geomagnetic studies. In: Dwivedi, S.B., et al. (Ed.), Recent Advances in Civil Engineering, pp. 66–73.

Rai, P.K., 2011b. Biomonitoring of particulates through magnetic properties of road-side plant leaves. In: Tiwari, D. (Ed.), Advances in Environmental Chemistry. Excel India Publishers, New Delhi, pp. 34–37.

Rai, P.K., 2013. Environmental magnetic studies of particulates with special reference to bio magnetic monitoring using roadside plant leaves. Atmospheric Environment 72, 113–129.

Rai, P.K., Mishra, A., Tripathi, B.D., 2010. Heavy metals and microbial pollution of river Ganga: a case study on water quality at Varanasi. Aquatic Ecosystem Health and Management 13 (4), 352–361.

Rai, P.K., Sharma, A.P., Tripathi, B.D., 2007. Urban environment status in Singrauli industrial region and its eco-sustainable management: a case study on heavy metal pollution. In: Lakshmi, V. (Ed.), Urban Planing and Environment, Strategies and Challenges. McMillan Advanced Research Series, pp. 213–217.

Rajput, M., Agrawal, M., 2005. Biomonitoring of air pollution in a seasonally dry tropical suburban area using wheat transplants. Environment Monitoring and Assessment 101, 39–53.

Rautio, P., Huttunen, S., Lamppu, J., 1998. Element concentrations in Scots pine needles on radial transects across a subarctic area. Water, Air and Soil Pollution 102, 389–405.

Rea, D.K., Hovan, S.A., 1995. Grain-size distribution and depositional processes of the mineral component of Abyssal sediments – lessons from the North Pacific. Paleoceanography 10, 251–258.

Retallack, G.J., 2001. Soils of the Past: An Introduction to Paleopedology, second ed. Blackwell, Oxford 600.

Revuelta, M.A., et al., 2014. Partitioning of magnetic particles in PM_{10}, $PM_{2.5}$ and PM_1 aerosols in the urban atmosphere of Barcelona (Spain). Environmental Pollution 188, 109–117.

Risom, L., Møller, P., Loft, S., 2005. Oxidative stress-induced DNA damage by particulate air pollution. Mutation Research 592, 119–137.

Rizzio, E., Giaveri, G., Arginelli, D., Gini, L., Profumo, A., Gallorini, M., 1999. Trace elements total content and particle sizes distribution in the air particulate matter of a rural-residential area in the north Italy investigated by instrumental neutron activation analysis. Science of the Total Environment 226, 47–56.

Robertson, D.J., Taylor, K.G., Hoon, S.R., 2003. Geochemical and mineral magnetic characterisation of urban sediment particulates, Manchester, UK. Applied Geochemistry 18 (2), 269–282.

Rodríguez-Germade, I., Mohamed, K.J., Rubio, B., García, A., 2014. The influence of weather and climate on the reliability of magnetic properties of tree leaves as proxies for air pollution monitoring. Science of the Total Environment 468–469, 892–902.

Rowe, P.J., Maher, B.A., 2000. 'Cold' stage formation of calcrete nodules in the Chinese Loess Plateau: evidence from U-series dating and stable isotope analysis. Palaeogeography Palaeoclimatology Palaeoecology 157, 109–125.

Sagnotti, L., Macrí, P., Egli, R., Mondolio, M., 2006. Magnetic properties of atmospheric particulate matter from automatic air sampler stations in Latium (Italy): towards a definition of magnetic fingerprints for natural and anthropogenic PM_{10} sources. Journal of Geophysical Research 111, B12S22.

Sagnotti, L., Taddeucci, J., Winkler, A., Cavallo, A., et al., 2009. Compositional, morphological, and hysteresis characterization of magnetic airborne particulate matter in Rome, Italy. Geochemistry Geophysics Geosystems 10, 17. http://dx.doi.org/10.1029/2009GC002563 Q08Z06.

Saldiva, P.H., Clarke, R.W., Coull, B.A., Stearns, R.C., Lawrence, J., Murthy, G.G., Diaz, E., Koutrakis, P., Suth, H., Tsuda, A., Godleski, J.J., 2002. Lung inflammation induced by concentrated ambient air particles is related to particle composition. American Journal of Respiratory and Critical Care Medicine 165, 1610–1617.

Samal, A.K., Santra, S.C., 2002. Indian Journal of Environmental Health 44, 71–76.

Saquib, M., Ahmad, Z., Zaheer, A., Malabar, A.A., 1992. Journal of Environmental Biology 13, 145–148.

Saragnese, F., Lanci, L., Lanza, R., 2011. Nanometric-sized atmospheric particulate studied by magnetic analyses. Atmospheric Environment 45, 450–459.

Sassen, K., DeMott, P.J., Prospero, J.M., Poellot, M.R., 2003. Saharan dust storms and indirect aerosol effects on clouds: CRYSTAL-FACE Results. Geophysical Research Letters 30. http://dx.doi.org/10.1029/2003GL017371.

Schadlich, G., Weissflog, L., Schuurmann, G., 1995. Magnetic susceptibility in conifer needles as indicator of fly ash deposition. Fresenius Environmental Bulletin 4, 7–12.

Schoket, B., 1999. DNA damage in humans exposed to environmental and dietary polycyclic aromatic hydrocarbons. Mutation Research 424, 143–153.

Schwartz, J., 1996. Air pollution and hospital admissions for respiratory disease. Epidemiology 7, 20–28.

Schwarze, P.E., Ovrevik, J., Lag, M., Refsnes, M., Nafstad, P., Hetland, R.B., et al., 2006. Particulate matter properties and health effects: consistency of epidemiological and toxicological studies. Human and Experimental Toxicology 25, 559–579.

Schwertmann, U., Taylor, R.M., 1977. Iron oxides in minerals. In: Dixon, J.B., Weed, S.B. (Eds.), Soil Environments. Soil Sci. Soc. Am, Madison, WI, pp. 145–180.

Seaton, A., MacNee, W., Donaldson, K., 1995. Particulate air pollution and acute health effects. Lancet 345, 176–178.

Senapati, M.R., Misra, P.K., 1996. Impact of auto exhaust lead pollution on vegetation. Pollution Research 15, 109–111.

Sharma, A., Tripathi, B.D., 2008. Magnetic mapping of fly-ash pollution and heavy metals from soil samples around a point source in a dry tropical environment. Environmental Monitoring and Assessment 138 (1–3), 31–39.

Sharma, A.P., Rai, P.K., Tripathi, B.D., 2007. Magnetic biomonitoring of roadside tree leaves as a proxy of vehicular pollution. In: Vyas, L. (Ed.), Urban Planning and Environment: Strategies and Challenges. Publisher Macmillan Advanced Research Series, pp. 326–331.

Shigemitsu, M., Narita, H., Watanabe, Y.W., Harada, N., Tsunogai, S., 2007. Ba, Si, U, Al, Sc, La, Th, C and C-13/C-12 in a sediment core in the western subarctic Pacific as proxies of past biological production. Marine Chemistry 106, 442–455.

Shilton, V.F., Booth, C.A., Smith, J.P., Giess, P., Mitchell, D.J., Williams, C.D., 2005. Magnetic properties of urban street dust and their relationship with organic matter content in the West Midlands, UK. Atmospheric Environment 39, 3651–3659.

Shrivastava, N., Joshi, S., 2002. Effect of automobile air pollution on the growth of some plants at Kota. Geobios 29, 281–282.

Shu, J., Dearing, J., Morse, A., Yu, L., Li, C., 2000. Magnetic properties of daily sampled total suspended particulates in Shanghai. Environmental Science and Technology 34, 2393–2400.

Shu, J., Dearing, J.A., Morse, A.P., Yu, L., Yuan, N., 2001. Determining the sources of atmospheric particles in Shanghai, China, from magnetic and geochemical properties. Atmospheric Environment 35, 2615–2625.

Simonich, S., Hites, R., 1994. Importance of vegetation in removing polycyclic hydrocarbons from the atmosphere. Nature 370, 49–51.

Simonich, S., Hites, R., 1995. Organic pollutant accumulation in vegetation. Environmental Science and Technology 29, 2905–2914.

Singh, J.S., Singh, K.P., Agrawal, M., 1991. Environmental degradation of the Obra/Renukoot/Singrauli Area, India and its impact on natural and derived ecosystems. The Environmentalist 11, 171–180.

Singh, P., Sthapak, J., 1999. Reduction in protein contents in a few plants as indicators of air pollution. Pollution Research 18, 281–283.

Singh, R.B., 2000. Impact of stone crusher dust pollution on tomato (*Lycopersicum esculantum*) in the Sonbhadra district of U.P. Journal of Environmental Pollution 7, 235–239.

Singh, R.B., Das, U.C., Prasad, B.B., Jha, S.K., 2002. Pollution Research 21, 13–16.

Sokolik, I.N., Toon, O.B., 1999. Incorporation of mineralogical composition into models of the radiative properties of mineral aerosol from UV to IR wavelengths. Journal of Geophysics Research 104, 9423–9444.

Somashekar, R.K., Ravikumar, R., Ramesh, A.M., 1999. Impact of granite mining on some plant species around quarries and crusher of Bangalore district. Pollution Research 18, 445–451.

Spassov, S., Egli, R., Heller, F., Nourgaliev, D.K., Hannam, J., 2004. Magnetic quantification of urban pollution sources in atmospheric particulate matter. Geophysical Journal International 159, 555–564.

Spracklen, D.V., et al., 2008. Contribution of particle formation to global cloud condensation nuclei concentrations. Geophysical Research Letters 35, L06808. http://dx.doi.org/10.1029/2007GL033038.

Steinnes, E., Lukina, N., Nikonov, V., Aamlid, D., Royset, O., 2000. A gradient study of 34 elements in the vicinity of a copper-nickel smelter in the Kola peninsula. Environmental Monitoring and Assessment 60, 71–80.

Sul, D., Oh, E., Im, H., Yang, M., Kim, C.-W., Lee, E., 2003. DNA damage in T- and B-lymphocytes and granulocytes in emission inspection and incineration workers exposed to polycyclic aromatic hydrocarbons. Mutation Research/Genetic toxicology and Environmental Mutagenesis 538 (1–2), 109–119.

Szönyi, M., Sagnotti, L., Hirt, A.M., 2008. A refined biomonitoring study of airborne particulate matter pollution in Rome, with magnetic measurements of Quercus Ilex tree leaves. Geophysical Journal 173, 127–141.

Thompson, R., Oldfield, F., 1986. Environmental Magnetism. George Allen and Unwin, London, 230 pp.

Tiwari, T.N., Patel, M.K., 1993. Indian Journal of Environmental Protection 13, 93–95.

Trivedi, M.L., Singh, R.S., 1995. Reduction in protein contents in a few plants as indicators of air pollution. Pollution Research 14, 269–273.

Ulrich, B., 1984. Effects of air pollution on forest ecosystems and waters: the principles demonstrated at a case study in Central Europe. Atmosphere Environment 18, 621–628.

Urbat, M., Lehndorff, E., Schwark, L., 2004. Biomonitoring of air quality in the Cologne conurbation using pine needles as a passive sampler—part 1: magnetic properties. Atmospheric Environment 38, 3781–3792.

Valavanidis, A., Fiotakis, K., Vlachogianni, T., 2008. Airborne particulate matter and human health: toxicological assessment and importance of size and composition of particles for oxidative damage and carcinogenic mechanisms. Journal of Environmental Science and Health. Part C 26 (4), 339–362.

Veijalainen, H., 1998. The applicability of peat and needle analysis in heavy metal deposition surveys. Water, Air and Soil Pollution 107, 367–391.

Vuković, G., Urošević, M.A., Tomašević, M., Samson, R., Popović, A., 2015. Biomagnetic monitoring of urban air pollution using moss bags (*Sphagnum girgensohnii*). Ecological Indicators 52, 40–47.

Vyas, D., Krishnayya, N.S.R., Bedi, S.J., 1991. Indian Journal of Environmental Health 33, 260–263.

Watkins, S.J., Maher, B.A., 2003. Magnetic characterisation of present-day deep-sea sediments and sources in the North Atlantic. Earth and Planetary Science Letters 214, 379–394.

Watkins, S.J., Maher, B.A., Bigg, G.R., 2007. Ocean circulation at the Last Glacial Maximum: a combined modeling and magnetic proxy-based study. Paleoceanography 22.

Wichmann, H.E., Peters, A., 2000. Epidemiological evidence of the effects of ultrafine particle exposure. Philosophical Transaction of the Royal Society of London A 358, 2751–2769.

Wolterbeek, B., 2002. Biomonitoring of trace element air pollution: principles, possibilities and perspectives. Environmental Pollution 120, 11–21.

Wright, V.P. (Ed.), 1986. Palaeosols: Their Recognition and Interpretation. Blackwell, Oxford, 315 pp.

Xia, D.S., Chen, F.H., Bloemendal, J., Liu, X.M., Yu, Y., Yang, L.P., 2008. Magnetic properties of urban dustfall in Lanzhou, China, and its environmental implications. Atmospheric Environment 42 (9), 2198–2207.

Xie, S., Dearing, J.A., Bloemendal, J., 2000. The organic matter content of street dust in Liverpool, UK, and its association with dust magnetic properties. Atmospheric Environment 34, 269–275.

Xie, S., Dearing, J.A., Boyle, J.F., Bloemendal, J., Morse, A.P., 2001. Association between magnetic properties and element concentrations of Liverpool street dust and its implications. Journal of Applied Geophysics 48, 83–92.

Zafar, M., 1985. Pollution Research 4, 59–61.

Zeger, S.L., Dominici, F., McDermott, A., Samet, J.M., 2008. Mortality in the medicare population and chronic exposure to fine particulate air pollution in urban centers (2000–2005). Environmental Health Perspectives 116 (12), 1614–1619.

Zhang, C., Huang, B., Piper, J.D.A., Luo, R., 2008. Biomonitoring of atmospheric particulate matter using magnetic properties of Salix matsudana tree ring cores. Science of the total Environment 393 (1), 177–190.

Zhang, C., Qiao, Q., Piper, J.D.A., Huang, B., 2011. Assessment of heavy metal pollution from a Fe-smelting plant in urban river sediments using environmental magnetic and geochemical methods. Environmental Pollution. http://dx.doi.org/10.1016/j.envpol.2011.04.006.

Zhang, W., Yu, L., Lu, M., Hutchinson, M., Feng, H., 2007. Magnetic approach to normalizing heavy metal concentrations for particle size effects in intertidal sediments in the Yangtze Estuary, China. Environmental Pollution 147, 238–244.

CHAPTER SIX

Case Studies on Biomagnetic Monitoring of Particulates through Two Tropical Plant Species

6.1 INTRODUCTION

As already mentioned in previous chapters, particulate matter (PM) is one of the most problematic air pollutants in view of its adverse impacts on human health. Currently, existing conventional technologies for monitoring of PM are not always feasible, which has paved the way for using a magnetic biomonitoring approach. It has been demonstrated that magnetic measurement is an important means in PM pollution study through plant leaves. Plant species are found to be effective biomonitors and may act as natural filters by trapping and retaining PM on their leaf surfaces. In the developing countries, air quality crisis is attributed especially to vehicular emissions, which contributes to 40–80% of total air pollution (Ghose et al., 2005). The urban population is mainly exposed to high levels of air pollution including metals from motor vehicle emissions, which is also the main source of fine and ultrafine particles (Sharma et al., 2006; Rai, 2013; Rai and Panda, 2014; Rai et al., 2014). Road traffic is considered to be the major source of environmental pollution in urban areas, whereas other sources may be anthropogenic activities like power plants, metallurgy, mining and dust originating from fragile rocks. The impact of vehicular pollution on human health in urban areas is at peak level as vehicle emissions are near the ground level where people live and work. Atmospheric pollutants exist in both gaseous and pollutant form (Bucko et al., 2011). Many studies have highlighted the importance of PM with an aerodynamic diameter of less than $10\,\mu m$ (PM_{10}), which, due to their small size, can penetrate deep into the human lung and cause cardiovascular diseases (Le Tertre et al., 2002; Janssen et al., 2005; Jerrett et al., 2005; Rai, 2011a,b, 2013; Rai et al., 2014). Alongside PM_{10} are further grain

Biomagnetic Monitoring of Particulate Matter
ISBN 978-0-12-805135-1
http://dx.doi.org/10.1016/B978-0-12-805135-1.00006-8

size divisions of $PM_{2.5}$ and $PM_{0.1}$ (2.5 μm and 0.1 μm, respectively, again relative to their aerodynamic diameters). These fine and ultrafine particulates have higher burdens of toxicity as they become coated with heavy metals and chemicals, which, when inhaled, can become absorbed into the body and may target specific organs (Morawska and Zhang, 2002; Englert, 2004; Rai, 2013).

In early research, biogenic ferrimagnets were also reported to be present in the organisms like termites (Maher, 1998a) and bacteria (Fassbinder et al., 1990). However, it is now well established through a series of research that urban PM may also contain magnetic particles (Hunt et al., 1984; Flanders, 1994; Morris et al., 1995; Matzka and Maher, 1999; Petrovsky and Ellwood, 1999; Pandey et al., 2005; Maher et al., 2008; Rai, 2013; Rai et al., 2014). Iron often occurs as an impurity in fossil fuels during industrial, domestic, or vehicle combustion; carbon and organic material are lost by oxidation and the iron forms a nonvolatile residue, often comprising glassy spherules. These spherules are magnetic, with easily measurable magnetization levels (Maher, 2009). Also, combustion-related particles in vehicles, via exhaust emissions and abrasion/corrosion of engine and/or vehicle body material, can generate nonspherical magnetite particles (Matzka and Maher, 1999; Pandey et al., 2005; Maher, 2009; Rai, 2013; Rai et al., 2014). Gautam et al. (2004) measured magnetic susceptibility of soils, sediments, and roadside materials, inside and outside the Kathmandu urban area and magnetomineralogical analyses as well as scanning electron microscopy on magnetic extracts, grain size fractions, or bulk samples of road dust and soils, suggest lithogenic magnetite-like minerals and anthropogenic magnetic spherules to be the dominant contributors to the magnetic susceptibility signal. Magnetic minerals derived from vehicular combustion are mainly maghemite and metallic iron grains having a size range of 0.1–0.7 μm (Pandey et al., 2005; Maher, 2009; Rai, 2013). This grain size is particularly dangerous to humans because of its ability to be inhaled into the lungs (Matzka and Maher, 1999; Maher, 2009; Rai et al., 2014).

In view of the above mentioned deleterious impacts of PM, it is quite obvious that there is a need to investigate feasible and ecosustainable green technologies. Although there are many conventional (physical and chemical) devices for assessment of air pollution, biomonitoring is an efficient tool in urban areas (Rai, 2013; Rai et al., 2014). Biological monitors are organisms that provide quantitative information on some aspects of the environment, such as how much of a pollutant is present. In this regard, the air cleansing capacity of urban trees presents an alternative approach to foster an integrated approach to the sustainable management of urban ecosystems (Rai, 2013). Lichens, bryophytes, or mosses and certain conifers have recently been proven to be a potent biomonitoring tool of air pollution (Rai, 2013). However, in urban and peri-urban regions, higher plants are mostly suitable for monitoring dust or PM pollution as lichens and mosses are often absent in the landscape (Faiz et al., 2009; Rai, 2013). Further, urban trees and shrubs planted in street canyons proved to be efficient dust-capturing tools (Moreno et al., 2003; Urbat et al., 2004; Rai and Panda, 2014; Rai et al., 2014). Spreading widely in urban area and easily collected, tree leaves could improve the scanning resolution in the spatial scale (Mitchell et al., 2010; Yin et al., 2013; Rai et al., 2014). With the quick, economical, sensitive, and nondestructive feature of environmental magnetism measurement, the magnetic properties of tree leaves as a proxy in monitoring and mapping of PM pollution has drawn increasing attention (Yin et al., 2013; Rai et al., 2014). Moreover, tree leaves are efficient passive pollution collectors, as they provide a large surface for particle deposition, a large number of samples and sampling sites, and require no protection from vandalism (Sant'Ovaia et al., 2012). Therefore, urban angiosperm trees offer positive biological, ecological, and aerodynamic effects in comparison to lower groups of plants (Moreno et al., 2003; Urbat et al., 2004; Rai, 2013; Rai et al., 2013, 2014).

Biomagnetic monitoring with urban roadside tree leaves is very recent area of interest in the field of PM pollution science. The concept of environmental magnetism as a proxy for atmospheric pollution levels has been reported by several researchers based on analysis of soils and street or roof dust (Hay et al., 1997; Maher, 1998b; Hoffmann et al., 1999; Shu et al., 2000; Xie et al., 2000, 2001; Hanesch et al., 2007; Jordanova et al., 2003), and vegetation samples including tree bark (Kletetschka et al., 2003; Urbat et al., 2004). However, a cascade of research has emphasized the use of plant leaves in monitoring the dust or PM (Matzka and Maher, 1999; Moreno et al., 2003; Jordanova et al., 2003; Urbat et al., 2004; Pandey et al., 2005; Maher et al., 2008; Maher, 2009; Kardel et al., 2011; Rai, 2013; Rai et al., 2014). Maher and her group were the leaders in performing a multitude of magnetic studies in relation to the environment, thus helping turn it into a specialized discipline, i.e., environmental geomagnetism (Matzka and Maher, 1999; Maher et al., 2008; Maher et al., 2010; Kardel et al., 2011; Rai et al., 2014). In view of this, magnetic biomonitoring studies of roadside plant leaves were performed in the Singrauli and Varanasi regions of India (Pandey et al., 2005; Prajapati et al., 2006; Sharma et al., 2007; Rai, 2013), in some cities of Portugal, and hilly areas of Nepal (Gautam et al., 2004, 2005), in addition to a series of pioneering studies in Europe led by B. A. Maher (e.g., Matzka and Maher, 1999; Maher, 2009; Hansard et al., 2011; Kardel et al., 2011; Rai et al., 2014).

The particles of dust that deposit from the atmosphere and accumulate along the roadside are called road dust particles and originate from the interaction of solid, liquid, and gaseous metals (Rai, 2013; Rai et al., 2014). Since the roadside vegetation obviously comes into direct contact with particulates, irrespective of the sources, it is quite obvious that their role needs to be investigated, particularly in context of the role of plant leaves (Faiz et al., 2009). However, the diversity of plants investigated for their biomagnetic monitoring potential is limited mostly to plants prevailing in temperate conditions; therefore, the field is ripe for investigation in the context of tropical plants. Moreover, advancements in instrumentation or methodology related to magnetic studies may also improve understanding of this research field, which is still in its earliest stages (Rai, 2013).

Rapid urbanization and continuously expanding population has been the major cause of increase in the number of vehicles and hence particulates in Aizawl district, which lies in an Indo–Burma hot spot region that is connected to the National Highway road (NH-54; Pushpak) going to the airport, Silchar, Shillong, and finally Guawahati, which harbors a substantial vehicular fleet. Therefore, vehicular pollution may be the primary contributor of particulates, specifically respirable suspended particulate matter (RSPM), having human health implications. One preliminary study in Aizawl (Lalrinpuii and Lalramnghinglova, 2008) recorded the level of suspended particulate matter (SPM) and RSPM above the permissible limit of National Ambient Air Quality Standards. Further, PM below the size of $10\,\mu m$ (PM_{10}) are specifically hazardous to human health (Saldiva et al., 2002), therefore their monitoring is pertinent.

Apart from vehicular dust generation, other anthropogenic sources are soil erosion, mining, and stone quarrying activities prevailing particularly in peri-urban and rural regions of Aizawl (Rai, 2011a). Furthermore, increasingly, airborne dust particles emitted from geologic media pose threats to human health and the environment (Faiz et al., 2009). Since the rocks of Aizawl are very fragile, the weathered rock dust may also be deposited on plant leaves. In India several research studies demonstrated significant correlation between magnetic parameter and PM (Pandey et al., 2005; Prajapati et al., 2006; Rai et al., 2014), however, they analyzed only one magnetic parameter, i.e., isothermal remanent magnetization (IRM), whereas in the present study we included three parameters: magnetic susceptibility, anhysteretic remanent magnetization (ARM), and saturation isothermal remanent magnetization (SIRM). Therefore, this study aims to investigate the magnetic properties of two roadside plant leaves—*Hibiscus rosa-sinensis* and *Mangifera indica* at four spatially distant sites in order to compare their capability to accumulate particulates and to establish the relationship between magnetic properties and ambient PM.

6.2 DESCRIPTION OF STUDY SITE

Mizoram ($21°56'-24°31'N$ and $92°16'-93°26'E$) is one of the eight states in northeast India (Figure 6.1), covering an area of $21,081\,km^2$. The Tropic of Cancer divides the state into two almost equal parts. The state borders Myanmar to the east and south, Bangladesh to the west, and the states of Assam, Manipur, and Tripura to the north. The altitude rises towards the Myanmar border. The state of the forest vegetation falls into three major categories: tropical wet evergreen forest, tropical semi-evergreen forest, and subtropical pine forest (Champion and Seth, 1968). The region lies within an Indo-Burma hot spot region (Rai, 2009). Aizawl district comes under the Indo-Burma hot spot region of North East India (Rai, 2009, 2012); here, highly diverse plant species having varying leaf morphology can be sampled for dust deposition and study of magnetic parameters. Mostly the diversity of tropical evergreen plants prevails along the roadsides of Aizawl district, and therefore, they can retain the pollutants throughout the year, thus, offering no seasonal constraint to research.

Aizawl ($21°58'-21°85'N$ and $90°30'-90°60'E$), the capital of the state, is $1132\,m$ above sea level (asl). The altitude in Aizawl district varies from 800 to $1200\,masl$. The climate of the area is typically monsoonal, with an annual average rainfall *ca.* $2350\,mm$. The area experiences distinct seasons. The ambient air temperature is normally in the range of $20-30\,°C$ in summer and $11-21\,°C$ in winter (Laltlanchhuang, 2006). It is well known that meteorological data may also affect the air pollutants including dust or particulate deposition, therefore, average meteorological data of the study area recorded during the study period are noted in Table 6.1.

The study was carried out in Aizawl district from four different sampling points:

Site 1. Durtlang: Durtlang is a connecting road between Mizoram and Assam and is one of the main and busy roads of the city with very high

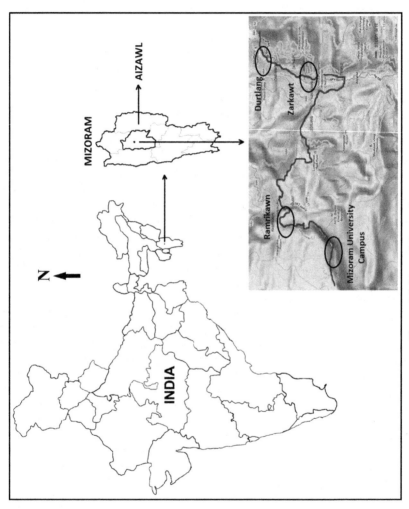

Figure 6.1 Map of the study area.

Table 6.1 Meteorological data of the study area (Rai et al., 2014)

Study period	Temperature		Rainfall (mm)	Humidity (%)
	Maximum °C	Minimum °C		
September, 2012	27.93	20.34	10.32	90.22
October, 2012	27.77	19.12	11.14	82.32
November, 2012	26.96	15.23	0	69.54
December, 2012	24.61	13.44	0	67.19
Average	**26.81**	**17.03**	**5.36**	**77.31**

traffic density. Vehicles are the main source of pollution at this site. *Site 2. Zarkawt*: Zarkawt is a commercial place in the city of Aizawl. Because of high traffic density, the emission of dust particles is usually very high in this area. *Site 3. Ramrikawn*: Ramrikawn is a very dense commercial area with markets, bus as well as taxi stands, and Food Corporation of India (FCI). FCI provides space for food storage for the entire Mizoram state. Due to the presence of FCI in Ramrikawn, there is frequent movement of heavy-duty vehicles coming from all parts of India through the National highway of Pushpak (NH-54). As there is a public bus and taxi stand, vehicular movement is usually high in Ramrikawn. Stone quarrying activity is also found in this area, which leads to emission of dust particles. Biomass burning through shifting cultivation is very common in this region (Rai, 2009, 2012) and may also be a source of suspended particulate matter pollution. In view of these pollution sources, we selected Ramrikawn as polluted area for investigation. *Site 4. Mizoram University Campus (MZU)*: MZU campus is an institutional area. Vehicles including buses, taxis, trucks, etc. are the main source of pollution in MZU campus. University buses, taxis, trucks, or trollies coming with construction materials are the main sources of pollution in MZU campus. However, the load of vehicles is very low and less frequent in comparison to other sites. Therefore, we selected MZU as a reference or control site in order to compare the results recorded from other sites. Further, we studied the winter season since dust or PM tend to concentrate during this season through atmospheric inversion (Verma and Singh, 2006) particularly during morning hours. In addition, in our recent research (Rai and Panda, 2014) we recorded maximum dust deposition during the winter season. This suggests that localized conditions like environmental, meteorological (Table 6.1), or anthropogenic may be influencing or disturbing particulate deposition or may reflect differences in the ability of leaf species to capture particulates (Power et al., 2009).

6.3 MATERIALS AND METHODS

Sampling was conducted during the months of September–December 2012 (a period of almost negligible rain as shown in Table 6.1). Tree leaves were collected from two species on dry sunny days: *H. rosa-sinensis* and *M. indica*. *Hibiscus rosa-sinensis* is an ornamental plant of Aizawl and *M. indica* is one of the most economically important tropical fruits in the state (it also has equally important medicinal value). From the literature we also found that several researchers (Prajapat et al., 2006; Rai et al., 2013) had used these plants species (*H. rosa-sinensis* and *M. indica*) because of their abundance, convenience for sampling, and their socioeconomic importance for the local people. Moreover, these plants had already been investigated for their suitability in efficient dust capturing (Rai et al., 2013; Rai and Panda, 2014). In addition, these plants are evergreen and therefore offer no seasonal constraints. At each site, five leaves of similar size from branches facing the roadside were plucked through random selection in the early hours of morning (8–12AM) and placed in polythene bags. Leaves were collected from the tree on the side nearest to the road at a height of approximately 2 m to avoid possible contamination from ground splash. Preference was usually given to the oldest leaves from the newest twig in order to select leaves of similar age and exposure time. The leaves were brought to the laboratory of the Department of Environmental Science, Mizoram University. Leaves were dried at 35 °C and the dried weight was recorded; samples were prepared for magnetic analysis, which involved packing the dried leaves into 10-cc plastic sample pots (Walden, 1999).

6.4 MAGNETIC PARAMETERS

The magnetic parameters such as magnetic susceptibility (χ), anhysteretic remanent magnetization, and saturation isothermal remanent magnetization were carried out with dried leaves in 10-cc plastic sample pots at K. S. Krishnan Geomagnetic Research Lab of the Indian Institute of Geomagnetism, Allahabad, Uttar Pradesh, India.

The magnetic susceptibility reflects the total composition of the dust deposited on the leaves, with a prevailing contribution from ferromagnetic minerals, which have much higher susceptibility values than paramagnetic and diamagnetic minerals, such as clay or quartz (Maher and Thompson, 1999; Evans and Heller, 2003; Sant'Ovaia et al., 2012). A Bartington (Oxford, England) MS-2B dual frequency susceptibility meter was used (Dearing, 1999) and measurements were taken. The sensitivity of this instrument was in the range of 10^{-6}.

Anhysteretic remanent magnetization indicates the magnetic concentration and is also sensitive to the presence of fine grains ~0.04–1 μm (Thompson and Oldfield, 1986), thus falling within the respirable size range of $PM_{2.5}$ and having the potential to have a high burden of toxicity (Power et al., 2009). This magnetization was induced in samples using a Molspin (Newcastle-upon-Tyne, England) AF demagnetizer, whereby a DC biasing field is generated in the presence of an alternating field, which peaks at 100 milli-Tesla (mT). The nature of this magnetic field magnetizes the fine magnetic grains, and the amount of magnetization retained within the sample (remanence) when removed from the field was measured using a Molspin1A magnetometer. The samples were then demagnetized to remove this induced field in preparation for the subsequent magnetic analysis (Walden, 1999).

Saturation isothermal remanent magnetization indicates the total concentration of magnetic grains (Evans and Heller, 2003) and can be used as a proxy of PM concentration (Muxworthy et al., 2003). This magnetization involves measuring the magnetic remanence of samples once removed from an induced field. Using a Molspin pulse magnetizer, a SIRM of 800 mT in the forward field was induced in the samples. At this high magnetization field, all magnetic grains within the sample become magnetized (Power et al., 2009). The instruments used for ARM and SIRM were fully automated.

The ratio of IRM_{-300} and SIRM was defined as the S-ratio (King and Channell, 1991). The S-ratio mainly reflects the relative proportion of antiferromagnetic to ferrimagnetic minerals in a sample. A ratio close to 1.0 reflects almost pure magnetite while ratios of <0.8 indicate the presence of some antiferromagnetic minerals, generally goethite or hematite (Thompson, 1986).

6.5 SUSPENDED PARTICULATE MATTER AND RESPIRABLE SUSPENDED PARTICULATE MATTER MONITORING

Sampling was done using a high volume sampler (Envirotech APM 460) 8 h daily for SPM and RSPM in the months of September–December, 2012 with a frequency of once in a week. The SPM in the atmosphere was determined using high volume method and the RSPM in the ambient air was determined using the cyclonic flow technique.

6.5.1 Heavy Metal (Fe)

Plant leaf samples were oven dried at 80 °C for 48 h and digested with aqua-regia and analyzed with an atomic absorption spectrophotometer.

6.6 STATISTICAL ANALYSIS

Correlation coefficient values were calculated at each site using SPSS software (SPSS Inc., version 10.0) to evaluate the relationship between PM and magnetic properties of *H. rosa-sinensis* and *M. indica* tree leaves, in order to assess this method as a proxy for particulate pollution and the suitability of leaves as depositories of particulate pollution.

6.7 RESULTS AND DISCUSSION

The ambient PM concentration recorded at spatially distant sites is shown in Table 6.2. The ambient PM concentrations were recorded highest at Ramrikawn, followed by Zarkawt and Durtlang, while the lowest values were recorded for MZU campus. The average magnetic data collected throughout the four-month sampling period is presented in Tables 6.3 and 6.4, respectively, for both *H. rosa-sinensis* and *M. indica* tree leaves (Rai et al., 2014).

In Durtlang, the χ value of *H. rosa-sinensis* was 16.12 ± 0.23 ($10^{-7}\,m^3\,kg^{-1}$), ARM was 14.47 ± 0.38 ($10^{-5}\,Am^2\,kg^{-1}$), and SIRM was 192.77 ± 0.11 ($10^{-5}\,Am^2\,kg^{-1}$). Similarly *M. indica* had a value of 17.19 ± 0.07 ($10^{-7}\,m^3\,kg^{-1}$) for χ, 15.23 ± 0.04 ($10^{-5}\,Am^2\,kg^{-1}$) for ARM, and 201.57 ± 0.21 ($10^{-5}\,Am^2\,kg^{-1}$) for SIRM (Rai et al., 2014).

In Zarkawt, χ, ARM, and SIRM values were 22.07 ± 0.39 ($10^{-7}\,m^3\,kg^{-1}$), 20.12 ± 0.21 ($10^{-5}\,Am^2\,kg^{-1}$), and SIRM 238.72 ± 0.28 ($10^{-5}\,Am^2\,kg^{-1}$), respectively, for *H. rosa-sinensis*. For *M. indica*, χ value was 26.14 ± 0.18 ($10^{-7}\,m^3\,kg^{-1}$), ARM value was 24.69 ± 0.08 ($10^{-5}\,Am^2\,kg^{-1}$), and SIRM value was 256.09 ± 0.29 ($10^{-5}\,Am^2\,kg^{-1}$) (Rai et al., 2014).

Table 6.2 Table showing the average suspended particulate matter (SPM) and respirable suspended particulate matter (RSPM) recorded from different sites during the study period (Rai et al., 2014)

Sampling location	SPM ($\mu g\,m^{-3}$)	RSPM ($\mu g\,m^{-3}$)
Durtlang	199.04	172.71
Zarkawt	219.13	190.09
Ramrikawn	250.07	220.12
MZU campus	130.12	100.09

Table 6.3 Summary of the magnetic data (mean and standard deviation) for roadside dust on *Hibiscus rosa-sinensis* tree leaves in the different sampling sites (Rai et al., 2014)

Site	χ (10^{-7} m³ kg⁻¹)	ARM (10^{-5} Am² kg⁻¹)	SIRM (10^{-5} Am² kg⁻¹)	ARM/χ (10^2 Am⁻¹)	SIRM/χ (10^2 Am⁻¹)	S-ratio
Durtlang	16.12 ± 0.23	14.47 ± 0.38	192.77 ± 0.11	0.89	11.95	0.931
Zarkawt	22.07 ± 0.39	20.12 ± 0.21	238.72 ± 0.28	0.91	10.81	0.901
Ramrikawn	25.12 ± 0.38	23.72 ± 0.22	266.19 ± 0.36	0.94	10.59	0.871
MZU campus	11.14 ± 0.09	9.38 ± 0.11	150.44 ± 0.15	0.84	13.50	0.873

Table 6.4 Summary of the magnetic data (mean and standard deviation) for roadside dust on *Mangifera indica* tree leaves in the different sampling sites (Rai et al., 2014)

Site	χ (10^{-7} m³ kg⁻¹)	ARM (10^{-5} Am² kg⁻¹)	SIRM (10^{-5} Am² kg⁻¹)	ARM/χ (10^2 Am⁻¹)	SIRM/χ (10^2 Am⁻¹)	S-ratio
Durtlang	17.19 ± 0.07	15.23 ± 0.04	201.57 ± 0.21	0.88	11.72	0.925
Zarkawt	26.14 ± 0.18	24.69 ± 0.08	256.09 ± 0.29	0.94	9.79	0.972
Ramrikawn	28.42 ± 0.08	29.32 ± 0.22	273.41 ± 0.31	1.03	9.62	0.938
MZU campus	12.13 ± 0.11	10.18 ± 0.18	168.76 ± 0.18	0.83	13.91	0.981

In Ramrikawn, the χ value of *H. rosa-sinensis* was 25.12 ± 0.38 ($10^{-7}\,m^3\,kg^{-1}$), ARM was 23.72 ± 0.22 ($10^{-5}\,Am^2\,kg^{-1}$), and SIRM was 266.19 ± 0.36 ($10^{-5}\,Am^2\,kg^{-1}$). Similarly, *M. indica* had an χ value of 28.42 ± 0.08 ($10^{-7}\,m^3\,kg^{-1}$), ARM value of 29.32 ± 0.22 ($10^{-5}\,Am^2\,kg^{-1}$), and SIRM value of 273.41 ± 0.31 ($10^{-5}\,Am^2\,kg^{-1}$) (Rai et al., 2014).

In MZU campus, the χ, ARM, and SIRM values were 11.14 ± 0.09 ($10^{-7}\,m^3\,kg^{-1}$), 9.38 ± 0.11 ($10^{-5}\,Am^2\,kg^{-1}$), and 150.44 ± 0.15 ($10^{-5}\,Am^2\,kg^{-1}$), respectively for *H. rosa-sinensis*. For *M. indica*, χ value was 12.13 ± 0.11 ($10^{-7}\,m^3\,kg^{-1}$), ARM value was 10.18 ± 0.18 ($10^{-5}\,Am^2\,kg^{-1}$), and SIRM value was 168.76 ± 0.18 ($10^{-5}\,Am^2\,kg^{-1}$). Several studies have demonstrated that magnetic susceptibility has been used as a proxy to monitor the regional distribution of air PM pollution or relative changes in an area (e.g., Moreno et al., 2003; Gautam et al., 2005; Sant'Ovaia et al., 2012; Rai et al., 2014).

The values of ARM/χ and SIRM/χ can reflect the grain size of magnetic minerals (Thompson and Oldfield, 1986; Evans and Heler, 2003). From the study, it was observed that ARM/χ and SIRM/χ values were low at all study sites (Tables 6.3 and 6.4). Low values of ARM/χ and SIRM/χ indicate relatively large grain size magnetic particles in leaf samples (Yin et al., 2013). S-ratio of leaf samples ranged from 0.871 to 0.981 (Tables 6.3 and 6.4), with an average of 0.924, which means that these leaf samples are dominated by "soft" magnetic minerals with a low coercive force, but a minor part of "hard" magnetic minerals with a relatively high coercive force also exists (Robinson, 1986). Muxworthy et al. (2003) advocated that SIRM was found to be strongly correlated with the PM mass and not only acts as a proxy for PM monitoring but also is a viable alternative to magnetic susceptibility when the samples are magnetically too weak.

From the findings recorded in Tables 6.3 and 6.4 we can infer that the magnetic values for both species display similar trends, with Ramrikawn representing the highest and MZU campus representing the lowest concentration data. Further, results indicate that Ramrikawn and Zarkawt experienced the highest deposition of magnetic grains, originating from PM. The χ, ARM, and SIRM values were high for *M. indica* when compared with *H. rosa-sinensis*. However, spatial trends of all three magnetic parameters displayed a similar trend with Ramrikawn, having maximum value and MZU campus recording the lowest value. The correlation coefficients indicated a significant relationship between the concentration of PM and magnetic measurement of both the roadside plant leaves (Tables 6.5 and 6.6; see Figures 6.2–6.13). In the literature, Hansard et al. (2011) also studied atmospheric particle pollution emitted by a combustion plant with tree leaves,

Table 6.5 Correlation between magnetic measurements (*Hibiscus rosa-sinensis*) with SPM and RSPM

Magnetic parameter	SPM (R^2)	RSPM (R^2)
χ	0.930	0.914
ARM	0.936	0.919
SIRM	0.939	0.923

Table 6.6 Correlation between magnetic measurements (*Mangifera indica*) with SPM and RSPM

Magnetic Parameter	SPM (R^2)	RSPM (R^2)
χ	0.882	0.863
ARM	0.879	0.857
SIRM	0.892	0.873

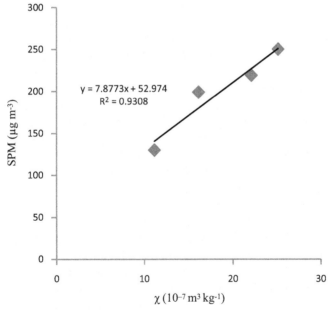

Figure 6.2 Correlation between χ (*Hibiscus rosa-sinensis*) and SPM.

and found that SIRM of leaf samples had a significant correlation with PM_{10} collected by a particle collector. Hu et al. (2008) also observed significant correlation of magnetic parameters (magnetic susceptibility, ARM and SIRM) with air pollutants, particularly heavy metals. Furthermore, Kardel et al. (2011) also recorded significant correlation between leaf SIRM and

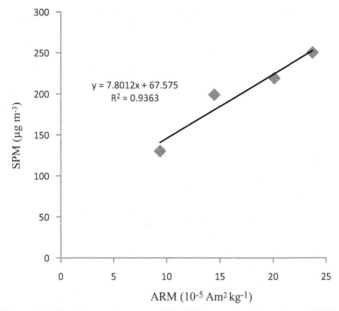

Figure 6.3 Correlation between ARM (*Hibiscus rosa-sinensis*) and SPM.

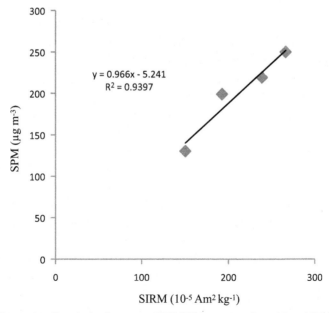

Figure 6.4 Correlation between SIRM (*Hibiscus rosa-sinensis*) and SPM.

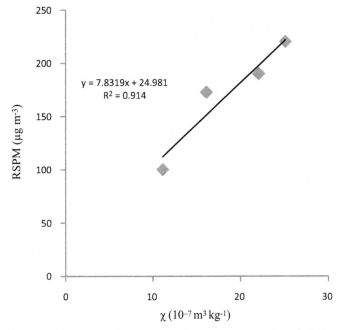

Figure 6.5 Correlation between χ (*Hibiscus rosa-sinensis*) and RSPM.

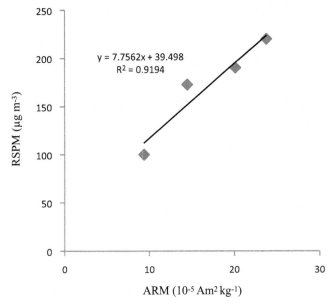

Figure 6.6 Correlation between ARM (*Hibiscus rosa-sinensis*) and RSPM.

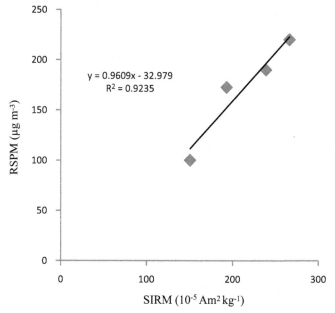

Figure 6.7 Correlation between SIRM (*Hibiscus rosa-sinensis*) and RSPM.

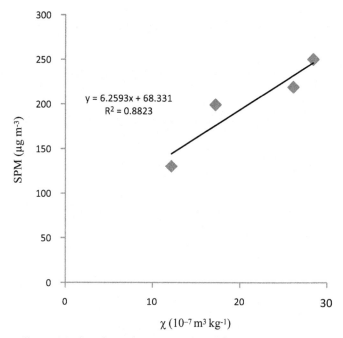

Figure 6.8 Correlation between χ (*Mangifera indica*) and SPM.

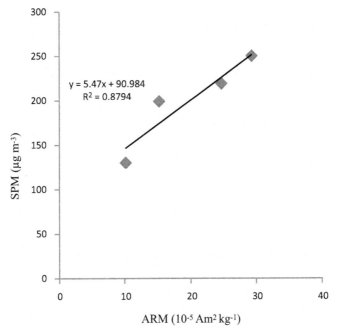

Figure 6.9 Correlation between ARM (*Mangifera indica*) and SPM.

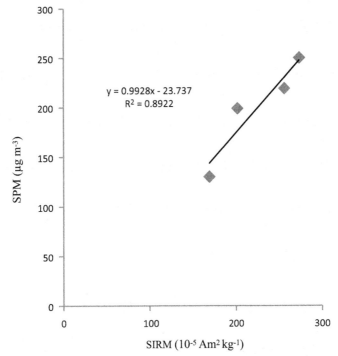

Figure 6.10 Correlation between SIRM (*Mangifera indica*) and SPM.

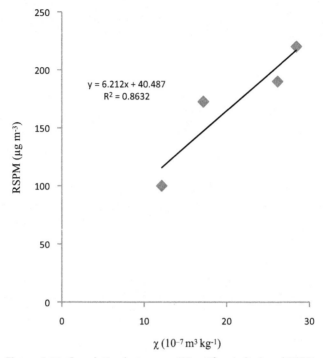

Figure 6.11 Correlation between χ (*Mangifera indica*) and RSPM.

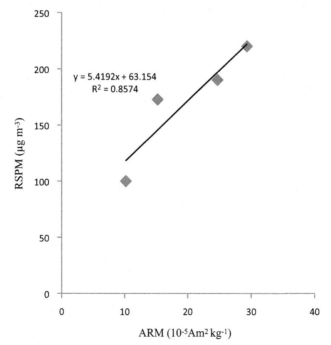

Figure 6.12 Correlation between ARM (*Mangifera indica*) and RSPM.

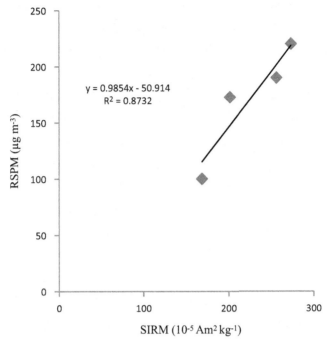

Figure 6.13 Correlation between SIRM (*Mangifera indica*) and RSPM.

ambient PM concentrations. Also in India, several studies demonstrated a significant correlation between magnetic parameter and PM (Pandey et al., 2005; Prajapati et al., 2006).

The average magnetic concentration data (Tables 6.3 and 6.4) demonstrate that the accumulation of PM on tree leaves varies across the four locations. The results suggest that Ramrikawn and Zarkawt experience the heaviest loads of particulates in comparison to the low-deposition sites at Durtlang and MZU campus. Ramrikawn recorded the highest values of magnetic parameters, which may be attributed to heavy vehicle load (due to location of the Food Corporation of India), street dust, and dust from fragile rocks. Zarkawt and Durtlang may have vehicular pollution as the only source of PM while MZU campus, being an institutional area, is relatively free from vehicular pollution and other anthropogenic activities.

Significant correlation coefficients have been recorded between PM and magnetic parameters of plant leaves, which indicated that roadside dust was comprised of magnetic particles (Figure 6.1–6.13). It was also observed that all magnetic parameters showed significant correlation with Fe (Table 6.7; see Figure 6.2 and Figures 6.14–6.19). Sant'Ovaia et al. (2012) also demonstrated

Table 6.7 Concentration of Fe (mg kg^{-1}) of the leaf samples
Mean concentration of Fe (mg kg^{-1}) ± S.D

Sampling site	*Hibiscus rosa-sinensis*	*Mangifera indica*
Durtlang	14.9 ± 0.42	16.23 ± 0.41
Zarkawt	18.01 ± 0.32	22.51 ± 0.21
Ramrikawn	18.82 ± 0.21	24.51 ± 0.58
MZU campus	12.81 ± 0.61	14.17 ± 0.81

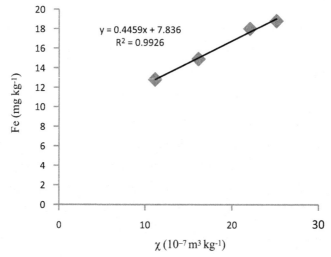

Figure 6.14 Correlation between Fe and χ (*Hibiscus rosa-sinensis*).

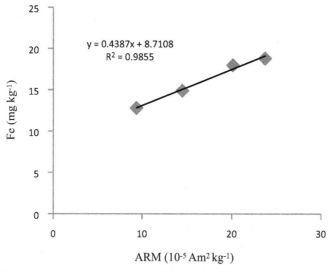

Figure 6.15 Correlation between Fe and ARM (*Hibiscus rosa-sinensis*).

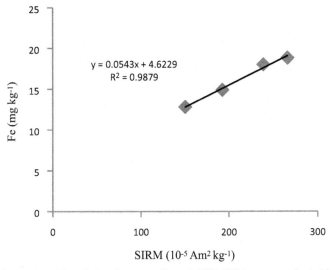

Figure 6.16 Correlation between Fe and SIRM (*Hibiscus rosa-sinensis*).

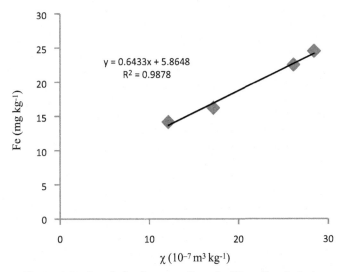

Figure 6.17 Correlation between Fe and χ (*Mangifera indica*).

positive significant correlation of magnetic parameters (magnetic suscep-tibility and SIRM) with Fe. Furthermore, strong correlation between the magnetic susceptibility of pine needles and their metal (Fe) content has been demonstrated due to deposition of fly ash particles (Maher et al., 2008). The observations indicated that magnetic properties of dust-loaded particles act as a proxy for ambient PM pollution levels (Rai et al., 2014).

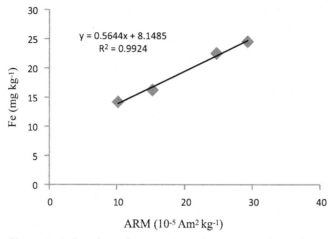

Figure 6.18 Correlation between Fe and ARM (*Mangifera indica*).

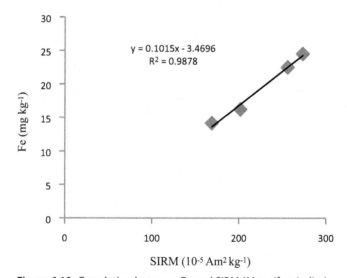

Figure 6.19 Correlation between Fe and SIRM (*Mangifera indica*).

6.8 CONCLUSIONS

Results from this work validate the magnetic analysis of roadside tree leaves as proxy indicators of PM pollution. Magnetic concentration data suggest that the deposition of PM on roadside tree leaves varies due to varying traffic behavior between sites and due to other activities like soil erosion, mining, and stone quarrying, etc. The magnetic analysis of dust loadings on roadside tree leaves provides an alternative proxy method to conventional pollution monitoring. Further, the present study is a strong first step and warrants further efforts that may pave the way to screen the feasibility of these plants in context of their potential to be planted in other urban areas with varying pollution load. In short, biomagnetic monitoring of PM through plant leaves provides a rapid and economical technique for monitoring atmospheric PM pollution, thus paving the way for innovation of an ecosustainable environmental management tool (Rai et al., 2014).

REFERENCES

Bucko, M.S., Magiera, T., Johanson, B., Petrovsky, E., Pesonen, L.J., 2011. Identification of magnetic particulates in road dust accumulated on roadside snow using magnetic, geochemical and micro-morphological analysis. Environmental Pollution 159, 1266–1276.

Champion, H., Seth, S.K., 1968. A Revised Survey of Forest Types of India. Government of India Press, Delhi.

Dearing, J., 1999. Magnetic susceptibility. Technical guide no. 6. In: Walden, J., Oldfield, F., Smith, J.P. (Eds.), Environmental Magnetism: A Practical Guide. Quaternary Research Association, Cambridge, England, pp. 35–62.

Englert, N., 2004. Fine particles and human health—a review of epidemiological studies. Toxicology Letters 149, 235–242.

Evans, M.E., Heller, F., 2003. Environmental Magnetism: Principles and Applications of Enviromagnetics (International Geophysics). Academic Press, Elsevier, London. 299 pp.

Faiz, Y., Tufail, M., Javed, M.T., Chaudhry, M.M., Siddique, N., 2009. Road dust pollution of Cd, Cu, Ni, Pb and Zn along Islamabad Expressway, Pakistan. Microchemical Journal 92, 186–192.

Fassbinder, J.W.E., Stanjek, H., Vali, H., 1990. Occurrence of magnetic bacteria in soil. Nature 343, 161–163.

Flanders, P.J., 1994. Collection, measurement, and analysis of airborne magnetic particulates from pollution in the environment. Journal of Applied Physics 75, 5931–5936.

Gautam, P., Blaha, U., Appel, E., Neupane, G., 2004. Environmental magnetic approach towards the quantification of pollution in Kathmandu urban area, Nepal. Physics and Chemistry of the Earth, Parts A/B/C 29 (13–14), 973–984.

Gautam, P., Blaha, U., Appel, E., 2005. Magnetic susceptibility of dust-loaded leaves as a proxy of traffic-related heavy metal pollution in Kathmandu city, Nepal. Atmospheric Environment 39, 2201–2211.

Ghose, M.K., Paul, R., Banerjee, R.K., 2005. Assessment of the status of urban air pollution and its impact on human health in the city of Kolkata. Environmental Monitoring and Assessment 108, 151–167.

Hanesch, M., Rantitsch, G., Hemetsberger, S., Scholger, R., 2007. Lithological and pedological influences on the magnetic susceptibility of soil: their consideration in magnetic pollution mapping. Geophysical Journal International 382 (23), 351–363.

Hansard, R., Maher, B.A., Kinnersley, R., 2011. Biomagnetic monitoring of industry-derived particulate pollution. Environmental Pollution 159 (6), 1673–1681.

Hay, K.L., Dearing, J.A., Baban, S.M.J., Loveland, P., 1997. A preliminary attempt to identify atmospherically derived pollution particles in English topsoils from magnetic susceptibility measurements. Physics and Chemistry of the Earth 22, 207–210.

Hoffmann, V., Knab, M., Appel, E., 1999. Magnetic susceptibility mapping of roadside pollution. Journal of Geochemical Exploration 66, 313–326.

Hu, S.Y., Duan, X.M., Shen, M.J., et al., 2008. Magnetic response to atmospheric heavy metal.pollution recorded by dust-loaded leaves in Shougang industrial area, western Beijing. Chinese Science Bulletin 53 (10), 1555–1564.

Hunt, A., Jones, J., Oldfield, F., 1984. Magnetic measurements and heavy metals in atmospheric particulates of anthropogenic origin. Science of the Total Environment 33, 129–139.

Janssen, N.A.H., Lanki, T., Hoek, G., Vallius, M., de Hartog, J.J., Van Grieken, R., 2005. Associations between ambient, personal and indoor exposure to fine particulate matter constituents in Dutch and Finnish panels of cardiovascular patients. Occupational and Environmental Medicine 62, 868–877.

Jerrett, M., Buzzelli, M., Burnett, R.T., DeLuca, P.F., 2005. Particulate air pollution, social confounders, and mortality in small areas of an industrial city. Social Science and Medicine 60, 2845–2863.

Jordanova, N.V., Jordanova, D.V., Veneva, L., Yorova, K., Petrovsky, E., 2003. Magnetic response of soils and vegetation to heavy metal pollution–a case study. Environmental Science and Technology 37, 4417–4424.

Kardel, F., Wuyts, K., Maher, B.A., Hansard, R., Samson, R., 2011. Leaf saturation isothermal remanent magnetization (SIRM) as a proxy for particulate matter monitoring: interspecies differences and in-season variation. Atmospheric Environment 45, 5164–5171.

King, J.W., Channell, J.E.T., 1991. Sedimentary magnetism, environmental magnetism and magnetostratigraphy. Reviews of Geophysics 29, 358–370.

Kletetschka, G., Zila, V., Wasilewski, P.J., 2003. Magnetic anomalies on the tree trunks. Studia Geophysica et Geodaetica 47, 371–379.

Lalrinpuii, H., Lalramnghinglova, H., 2008. Assessment of air pollution in Aizawl city. Current Science 94 (7), 852–853.

Laltlanchhuang, S.K., 2006. Studies of the Impact of Disturbance on Secondary Productivity of Forest Ecosystem with Special Reference to Surface, Sub-surface Litter Insect and Other Non-Insect Groups (M.Sc. dissertation). Mizoram University.

Le Tertre, A., Medina, S., Samoli, E., Forsberg, B., Michelozzi, P., Boumghar, A., 2002. Short-term effects of particulate air pollution on cardiovascular diseases in eight European cities. Journal of Epidemiology and Community Health 56, 773–779.

Maher, B.A., Thompson, R., 1999. Quaternary, Climates, Environments and Magnetism. Cambride University Press. 390 pp.

Maher, B.A., Mooreb, C., Matzka, J., 2008. Spatial variation in vehicle-derived metal pollution identified by magnetic and elemental analysis of roadside tree leaves. Atmospheric Environment 42, 364–373.

Maher, B.A., Mitchell, R., Kinnersley, R., 2010. High-resolution magnetic biomonitoring: a quantitative surrogate for particulate pollution. In: CLIMAQS Workshop 'Local Air Quality and its Interactions with Vegetation'.

Maher, B.A., 1998a. Magnetic biomineralisation in termites. Proceedings of the Royal Society of London 265, 233–237.

Maher, B.A., 1998b. Magnetic properties of modern soils and Quaternary loessicpaleosols: paleoclimatic implications. Palaeogeography, Palaeoclimatology, Palaeoecology 137, 25–54.

Maher, B.A., 2009. Rain and dust: magnetic records of climate and pollution. Elements 5, 229–234.

Matzka, J., Maher, B.A., 1999. Magnetic biomonitoring of roadside tree leaves:identification of spatial and temporal variations in vehicle-derived particulates. Atmospheric Environment 33, 4565–4569.

Mitchell, R., Maher, B.A., Kinnersley, R., 2010. Rates of particulate pollution deposition onto leaf surfaces: temporal and interspecies magnetic analyses. Environmental Pollution 158 (5), 1472–1478.

Morawska, L., Zhang, J., 2002. Combustion sources of particles. 1. Health relevance and source signatures. Chemosphere 49, 1045–1058.

Moreno, E., Sagnotti, L., Dinare_s-Turell, J., Winkler, A., Cascella, A., 2003. Biomonitoring of traffic air pollution in Rome using magnetic properties of tree leaves. Atmospheric Environment 37, 2967–2977.

Morris, W.A., Versteeg, J.K., Bryant, D.W., Legzdins, A.E., Mccarry, B.E., Marvin, C.H., 1995. Preliminary comparisons between mutagenicity and magnetic susceptibility of respirable airborne particulate. Atmospheric Environment 29, 3441–3450.

Muxworthy, A.R., Matzka, J., Davila, A.F., Petersen, N., 2003. Magnetic signature of daily sampled urban atmospheric particles. Atmospheric Environment 37, 4163–4169.

Pandey, S.K., Tripathi, B.D., Prajapati, S.K., Mishra, V.K., Upadhyay, A.R., Rai, P.K., Sharma, A.P., 2005. Magnetic properties of vehicle derived particulates and amelioration by *Ficus infectoria*: a keystone species. Ambio: A Journal on Human Environment 34 (8), 645–647.

Petrovsky, E., Ellwood, B.B., 1999. Magnetic monitoring of pollution of air, land andwaters. In: Maher, B.A., Thompson, R. (Eds.), Quaternary Climates, Environmentsand Magnetism. Cambridge University Press, Cambridge, pp. 279–322.

Power, A.L., Worsley, A.T., Booth, C., 2009. Magneto-biomonitoring of intra-urban spatial variations of particulate matter using tree leaves. Environmental Geochemistry and Health 31, 315–325.

Prajapati, S.K., Pandey, S.K., Tripathi, B.D., 2006. Magnetic biomonitoring of roadside tree leaves as a proxy of vehicular pollution. Environmental Monitoring and Assessment 120, 169–175.

Rai, P.K., Panda, L.S., Chutia, B.M., Singh, M.M., 2013. Comparative assessment of air pollution tolerance index (APTI) in the industrial (Rourkela) and non industrial area (Aizawl) of India: an eco-management approach. African Journal of Environmental Science and Technology 7 (10), 944–948.

Rai, P.K., Chutia, B.M., Patil, S.K., 2014. Monitoring of spatial variations of particulate matter (PM) pollution through biomagnetic aspects of roadside plant leaves in an Indo-Burma hot spot region. Urban Forestry & Urban Greening 13, 761–770.

Rai, P.K., Panda, L.S., 2014. Leaf dust deposition and its impact on biochemical aspect of some roadside plants in Aizawl, Mizoram, North-East India, International Research Journal of Environmental Sciences 3, 14–19.

Rai, P.K., 2009. Comparative assessment of soil properties after bamboo flowering and death in a tropical Forest of Indo-Burma hot spot. Ambio: A Journal on Human Environment 38 (2), 118–120.

Rai, P.K., 2011a. Dust deposition capacity of certain roadside plants in Aizawl, Mizoram: implications for environmental geomagnetic studies. In: Dwivedi, S.B., et al. (Ed.), Recent Advances in Civil Engineering, pp. 66–73.

Rai, P.K., 2011b. Biomonitoring of particulates through magnetic properties of road-side plant leaves. In: Tiwari, D. (Ed.), Advances in Environmental Chemistry. Excel India Publishers, New Delhi, pp. 34–37.

Rai, P.K., 2012. Assessment of multifaceted environmental issues and model development of an indo- Burma hot spot region. Environmental Monitoring and Assessment 184, 113–131.

Rai, P.K., 2013. Environmental magnetic studies of particulates with special reference to biomagnetic monitoring using roadside plant leaves. Atmospheric Environment 72, 113–129.

Robinson, S.G., 1986. The late Pleistocene palaeoclimatic record of North Atlantic deep-sea sediments revealed by mineral–magnetic measurements. Physics of the Earth and Planetary Interiors 42 (1–2), 22–47.

Saldiva, P.H., Clarke, R.W., Coull, B.A., Stearns, R.C., Lawrence, J., Murthy, G.G., Diaz, E., Koutrakis, P., Suth, H., Tsuda, A., Godleski, J.J., 2002. Lung inflammation induced by concentrated ambient air particles is related to particle composition. American Journal of Respiratory and Critical Care Medicine 165, 1610–1617.

Sant'Ovaia, H., Lacerda, M.J., Gomes, C., 2012. Particle pollution - an environmental magnetism study using biocollectors located in northern Portugal. Atmospheric Environment 61, 340–349.

Sharma, K., Singh, R., Barman, S.C., Mishra, D., Kumar, R., Negi, M.P.S., Mandal, S.K., Kisku, G.C., Khan, A.H., Kidwai, M.M., Bhargava, S.K., 2006. Comparison of trace metals concentration in PM_{10} of different location of Lucknow city, India. Bulletin of Environmental Contamination and Toxicology 77, 419–426.

Sharma, A.P., Rai, P.K., Tripathi, B.D., 2007. Magnetic biomonitoring of roadside tree leaves as a proxy of vehicular pollution. In: Lakshmi, V. (Ed.), Urban Planing and Environment: Strategies and Challenges. Mc Millan Advanced Research Series, pp. 326–331.

Shu, J., Dearing, J., Morse, A., Yu, L., Li, C., 2000. Magnetic properties of daily sampled total suspended particulates in Shanghai. Environmental Science and Technology 34, 2393–2400.

Thompson, R., Oldfield, F., 1986. Environmental Magnetism. Allen and Unwin, London. 227 pp.

Thompson, R., 1986. Modelling magnetization data using SIMPLEX. Phys. Earth planet. Inter 42, 113–127.

Urbat, M., Lehndorff, E., Schwark, L., 2004. Biomonitoring of air quality in the Cologne-conurbation using pine needles as a passive sampler part 1: magnetic properties. Atmospheric Environment 38, 3781–3792.

Verma, A., Singh, S.N., 2006. Biochemical and ultra structural changes in plant foliage exposed to auto pollution. Environmental Monitoring and Assessment 120, 585–602.

Walden, J., 1999. Sample collection and preparation. Technical guide no. 6. In: Walden, J., Oldfield, F., Smith, J.P. (Eds.), Environmental Magnetism: A Practical Guide. Quaternary Research Association, Cambridge, England, pp. 26–34.

Xie, S., Dearing, J.A., Bloemendal, J., 2000. The organic matter content of street dust in Liverpool, UK, and its association with dust magnetic properties. Atmospheric Environment 34, 269–275.

Xie, S., Dearing, J.A., Boyle, J.F., Bloemendal, J., Morse, A.P., 2001. Association between magnetic properties and element concentrations of Liverpool street dust and its implications. Journal of Applied Geophysics 48, 83–92.

Yin, G., Hu, S., Cao, L., Roesler, W., Appel, E., 2013. Magnetic properties of tree leaves and their significance in atmospheric particle pollution in Linfen city, China. Chinese Geographical Science 23 (1), 59–72.

CHAPTER SEVEN

Biomonitoring of Particulate Pollution Using Magnetic Properties of *Ficus bengalensis*: A Keystone Species

7.1 INTRODUCTION

Air pollution is one of the most serious problems faced by developing as well as developed countries in view of its adverse impacts on human health (Chauhan and Pawar, 2010). Atmospheric particulate matter (PM) is considered to be the largest contributor to the problem of urban air pollution (Szonyi et al., 2008). There is a strong correlation between PM and respiratory health damage (Schwartz, 1996; Pope et al., 2002; Knutsen et al., 2004; Knox, 2006; Yin et al., 2013; Rai, 2013; Rai et al., 2013; Rai and Panda, 2014; Rai and Chutia, 2015). Many studies highlight the importance of particulates with an aerodynamic diameter of less than $10\,\mu m$ (PM_{10}), which, due to their small size, can penetrate deep into the human lung and cause respiratory illness (Le Tertre et al., 2002; Janssen et al., 2005; Jerrett et al., 2005; Rai, 2013; Rai et al., 2013; Rai and Panda, 2014; Rai and Chutia, 2015). Alongside PM_{10} are further grain size divisions of $PM_{2.5}$ and $PM_{0.1}$ ($2.5\,\mu m$ and $0.1\,\mu m$, respectively, again relative to their aerodynamic diameters). These fine and ultrafine particulates have higher burdens of toxicity as they become coated with heavy metals and chemicals, which, when inhaled, can become absorbed into the body and may target specific organs (Morawska and Zhang, 2002; Englert, 2004; Power et al., 2009; Rai, 2013; Rai and Chutia, 2015). Urban anthropogenic PM contains certain heavy metals that will be toxic to human health (Harrison and Jones, 1995; Huhn et al., 1995; Rai and Chutia, 2015). Particulate matter is one of the most harmful pollution components widely present in the environment. It has been demonstrated that magnetic measurement is an important means in PM

Biomagnetic Monitoring of Particulate Matter
ISBN 978-0-12-805135-1
http://dx.doi.org/10.1016/B978-0-12-805135-1.00007-X

137

pollution study through plant leaves. Plants species are found as effective biomonitors and may act as natural filters by trapping and retaining PM on their surfaces.

In view of the abovementioned deleterious impacts of particulate matter, it is quite obvious that feasible and ecosustainable green technologies should be investigated. Although there are many conventional (physical and chemical) devices for assessment of air pollution, plant systems allow the direct assessment of air stressors particularly in the context of magnetic particles (Maher et al., 2010; Rai, 2013; Rai and Chutia, 2015). Biological monitors are organisms that provide quantitative information on some aspects of the environment, such as how much of a pollutant is present. In this regard, the air cleansing capacity of urban trees presents an alternative to foster an integrated approach to the sustainable management of the urban ecosystem (Rai, 2013). Moreover, in urban areas higher plants are mostly suitable for monitoring dust pollution as lichens and mosses are often missing (Faiz et al., 2009; Rai, 2013; Rai and Chutia, 2015).

Tree leaves have been proven to be good collectors of PM (Moreno et al., 2003; Urbat et al., 2004; Yin et al., 2013; Rai and Chutia, 2015). Biomonitoring of particulate pollution through magnetic properties of plant leaves is a reliable, rapid, and inexpensive alternative to conventional atmospheric pollution monitoring (Power et al., 2009; Rai and Chutia, 2015). This has promoted its suitability for aiding biomonitoring of air quality (Maher and Matzka, 1999; Moreno et al., 2003; Urbat et al., 2004; Rai and Chutia, 2015). The magnetic properties of tree leaves as a proxy in monitoring and mapping of PM pollution have drawn increasing attention (Zhang et al., 2006; Rai and Chutia, 2015).

In our research, we carried out a primary magnetic study on PM pollution in Aizawl city, Mizoram. The rapid urbanization, fast, drastic increases in vehicles on the roads and other activities including soil erosion, mining, stone quarrying, and shifting cultivation in Aizawl have led to increases in the concentration of particulate pollution in the atmosphere. In the present study, we collected *Ficus bengalensis* tree leaves from different parts of Aizawl city and conducted a series of environmental and particulate magnetic measurements, trying to map the PM pollution, to provide essential data for the recognition and control of air quality as well as for further environmental study (Rai and Chutia, 2015).

7.2 MATERIALS AND METHODS

7.2.1 Study Area

Mizoram (21°56′–24°31′N and 92°16′–93°26′E) is one of the eight states in northeast India (Figure 7.1), and it covers an area of 21,081 km². The Tropic of Cancer divides the state into two almost equal parts. The state is bordered with Myanmar to the east and south, Bangladesh to the west, and by the states of Assam, Manipur, and Tripura to the north. The altitude is approaching to near the Myanmar border. Mizoram's forest vegetation falls into three major categories: tropical wet evergreen forest, tropical semi-evergreen forest, and subtropical pine forest (Champion and Seth, 1968).

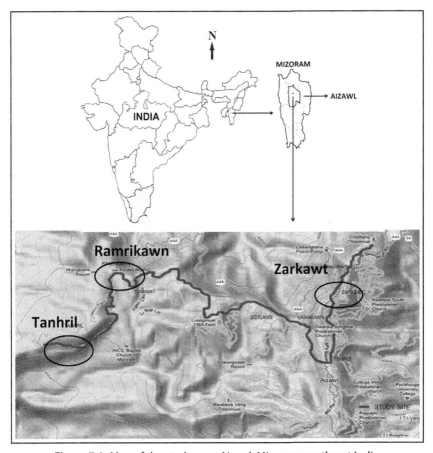

Figure 7.1 Map of the study area, Aizawl, Mizoram, northeast India.

Table 7.1 Meteorological data of the study area

Study period	Temperature (mm)		Rainfall	Humidity (%)
	Maximum °C	Minimum °C		
February, 2013	16.22	11.34	0	67.72
March, 2013	18.51	14.12	6.43	65.12
April, 2013	20.22	16.23	7.92	72.64
May, 2013	21.87	17.44	9.11	79.12
Average	**19.20**	**14.78**	**5.86**	**71.15**

Border Roads Task Force, Puspak, Aizawl.

Aizawl (21°58′–21°85′N and 90°30′–90°60′E), the capital of the state is 1132 m above sea level (asl). The altitude in Aizawl district varies from 800 to 1200 m asl. The climate of the area is typically monsoonal. The annual average rainfall is c. 2350 mm. The area experiences distinct seasons. The ambient air temperature normally is in the range of 20–30 °C in summer and 11–21 °C in winter (Laltlanchhuanga, 2006). Aizawl district comes under the Indo-Burma hotspot region of North East India (Rai and Panda, 2014). It is well known that meteorological data may also affect the air pollutants including dust or particulate deposition, therefore the average meteorological data of the study area recorded during the study period are cited in Table 7.1.

The study was carried out in Aizawl district, which was categorized into three subsites. *Site 1. Zarkawt:* Zarkawt is a commercial place in the city of Aizawl. Because of high traffic density, the emission of dust particles is very high. *Site 2. Ramrikawn:* Ramrikawn is a very densely located commercial area with markets, bus as well as taxi stands, and Food Corporation of India (FCI). FCI provides space for food storage for the entire Mizoram state. Due to the presence of FCI in Ramrikawn, there is frequent movement of heavy duty vehicles coming from all parts of India through National Highway of Pushpak (NH-54). As there is a public bus and taxi stand, vehicular movement is usually high in the Ramrikawn area. Stone quarrying activity is also found in this area, which leads to emission of dust particles. Biomass burning through shifting cultivation is very common in this region and may also be a source of suspended particulate matter pollution. In view of these pollution sources, we selected Ramrikawn as a polluted area for investigation. *Site 3. Tanhril:* It is a rural area having low vehicular activity, located in the western part of Aizawl district. However, the load of vehicles is very low and less frequent compared to other sites. Therefore, we selected

Tanhril as a reference or control site in order to compare the results recorded from other sites.

Sampling was conducted during the months of February, 2013 to May, 2013. Leaf samples were collected from *F. bengalensis* trees on dry sunny days. This plant was selected for sampling for the study because of its availability and commonness. Also, *Ficus* sp. is proved to be a potent biomagnetic monitoring tool (Rai, 2013). It also acts as a keystone species and its removal may lead to an extinction cascade. Also, this plant is evergreen, and therefore offers no seasonal constraint. At each site, 5–10 leaves of similar size from branches facing the roadside were plucked through random selection in the early hours of the morning (8–12 AM) and placed in polythene bags. Leaves were collected from the tree on the side nearest to the road at a height of approximately 2 m to avoid possible contamination from ground splash. The leaves were dried at 35 °C and the dried weights were recorded; samples were prepared for magnetic analysis, which involved packing the dried leaves into 10-cc plastic sample pots (Walden, 1999).

7.2.2 Magnetic Parameters

The magnetic parameters such as magnetic susceptibility (χ), anhysteretic remanent magnetization (ARM), and saturation isothermal remanent magnetization (SIRM) were carried out at K.S. Krishnan Geomagnetic Research Lab of the Indian Institute of Geomagnetism, Allahabad, India.

The magnetic susceptibility reflects the total composition of the dust deposited on the leaves, with a prevailing contribution from ferromagnetic minerals, which have much higher susceptibility values than paramagnetic and diamagnetic minerals, such as clay or quartz (Maher and Thompson, 1999; Evans and Heller, 2003). A Bartington (Oxford, England) MS2B dual frequency susceptibility meter was used (Dearing, 1999) and measurements were taken. The sensitivity of this instrument was in the range of 10^{-6}.

Anhysteretic remanent magnetization indicates the magnetic concentration and is also sensitive to the presence of fine grains ~0.04–1 μm (Thompson and Oldfield, 1986), thus falling within the respirable size range of $PM_{2.5}$ and having the potential to have a high burden of toxicity (Power et al., 2009). The ARM was induced in samples using a Molspin (Newcastle-upon-Tyne, England) AF demagnetizer, whereby a direct current biasing field is generated in the presence of an alternating field, which peaks at 100 milli-Tesla (mT). The nature of this magnetic field magnetizes the fine magnetic grains and the amount of magnetization retained within the sample (remanence) when removed from the field was measured using a Molspin1A magnetometer. The samples were then demagnetized to remove this induced field in preparation for the subsequent magnetic analysis (Walden, 1999).

Saturation isothermal remanent magnetization indicates the total concentration of magnetic grains (Evans and Heller, 2003) and can be used as a proxy of particulate matter concentration (Muxworthy et al., 2003). It involves measuring the magnetic remanence of samples once removed from an induced field. Using a Molspin pulse magnetizer, a SIRM of 800 mT in the forward field was induced in the samples. At this high magnetization, all magnetic grains within the sample become magnetized (Power et al., 2009). The instruments used for ARM and SIRM were fully automated.

The ratio of IRM-300 and SIRM was defined as the S-ratio (King and Channell, 1991). It mainly reflects the relative proportion of antiferromagnetic to ferrimagnetic minerals in a sample. A ratio close to 1.0 reflects almost pure magnetite while ratios of <0.8 indicate the presence of some antiferromagnetic minerals, generally goethite or hematite (Thompson, 1986).

7.2.3 Heavy Metals (Fe, Pb, and Cu)

Oven-dried *F. bengalensis* leaf samples were digested with aqua-regia and analyzed with an atomic absorption spectrophotometer for Fe, Pb, and Cu contents.

7.2.4 Statistical Analysis

Correlation coefficient values were calculated at each site using SPSS software (SPSS Inc., version 10.0).

7.3 RESULTS AND DISCUSSION

The highest ambient PM concentrations were recorded at Zarkawt, followed by Ramrikawn, while the lowest values were recorded for the Tanhril area. For *F. bengalensis* tree leaves, the average magnetic data collected throughout the four-month sampling period is presented in Table 7.2.

In Zarkawt, it was found that the χ value of *F. bengalensis* was 65.51 ± 0.42 ($10^{-7}\,\mathrm{m^3\,kg^{-1}}$), ARM was 26.53 ± 0.34 ($10^{-5}\,\mathrm{Am^2\,kg^{-1}}$), and SIRM was 498.81 ± 0.52 ($10^{-5}\,\mathrm{Am^2\,kg^{-1}}$). In Tanhril, the χ,

Table 7.2 Statistics of magnetic properties (mean and standard deviation) of *Ficus bengalensis* leaf

Site	χ (10^{-7} m^3 kg^{-1})	ARM (10^{-5} Am^2 kg^{-1})	SIRM (10^{-5} Am^2 kg^{-1})	ARM/χ (10^2 Am^{-1})	SIRM/χ (10^2 Am^{-1})	S-ratio
Zarkawt	65.51 ± 0.42	26.53 ± 0.34	498.81 ± 0.52	0.41	7.61	0.971
Tanhril	28.57 ± 0.17	4.46 ± 0.54	153.11 ± 0.23	0.15	5.35	0.965
Ramrikawn	37.09 ± 0.33	8.24 ± 0.72	203.61 ± 0.41	0.22	5.49	0.955

χ, magnetic susceptibility; ARM, anhysteretic remanent magnetization; SIRM, saturation isothermal remanent magnetization; S-ratio, isothermal remanent magnetization obtained under a backfield of 0.3T (IRM-300) versus saturation isothermal remanent magnetization.

ARM, and SIRM values were 28.57 ± 0.17 ($10^{-7}\,\text{m}^3\,\text{kg}^{-1}$), 4.46 ± 0.54 ($10^{-5}\,\text{Am}^2\,\text{kg}^{-1}$), and 153.11 ± 0.23 ($10^{-5}\,\text{Am}^2\,\text{kg}^{-1}$), respectively for *F. bengalensis*. In Ramrikawn, it was found that the χ value of *F. bengalensis* was 37.09 ± 0.33 ($10^{-7}\,\text{m}^3\,\text{kg}^{-1}$), ARM was 8.24 ± 0.72 ($10^{-5}\,\text{Am}^2\,\text{kg}^{-1}$), and SIRM was 203.61 ± 0.41 ($10^{-5}\,\text{Am}^2\,\text{kg}^{-1}$).

The high dispersion degrees of susceptibility and remanent magnetism mainly result from the sampling sites in different functional areas. Samples collected in the rural area show low susceptibility and remanent magnetism, where tree leaves sampled in city and peri-urban areas show higher values. The correlation of magnetic susceptibility with ARM and SIRM are significant (Figures 7.2 and 7.3). The relatively high correlation indicates that the magnetic minerals with paramagnetism and superparamagnetism contribute slightly to the magnetism of tree leaves, and the major contributor may be ferromagnetic minerals (Yu et al., 1995; Sun et al., 1996). The values of ARM/χ and SIRM/χ can reflect the grain size of magnetic minerals (Thompson and Oldfield, 1986; Evans and Heler, 2003). From the study it was observed that ARM/χ and SIRM/χ values are low at all study sites (Table 7.2). Low values of ARM/χ and SIRM/χ indicate relatively large grain size magnetic particles in leaf samples (Yin et al., 2013). S-ratio of *F. bengalensis* leaf samples ranged from 0.955 to 0.971 (Table 7.2) for three different study sites, which means that these leaf samples are dominated by "soft" magnetic minerals with a low coercive force, but a minor part of "hard" magnetic minerals with a relatively high coercive force also exists (Robinson, 1986; Rai and Chutia, 2015).

From the findings recorded in Table 7.2, we can conclude that Zarkawt has the highest concentration of magnetic data while Tanhril area has the lowest concentration data. Further, results indicate that Zarkawt and Ramrikawn experience the highest deposition of magnetic grains originating from PM.

The average magnetic concentration data (Table 7.2) demonstrates that the accumulation of PM on tree leaves varies across the three locations. The results suggest that Zarkawt and Ramrikawn experience the heaviest loads of particulates in comparison to the low-deposition site Tanhril area. This suggests that localized conditions like environmental, meteorological, or anthropogenic may be influencing or disturbing particulate deposition or it may reflect differences in the ability of leaf species to capture particulates (Power et al., 2009). Zarkawt recorded the highest values of magnetic parameters, which may be attributed to heavy vehicle load (due to city area) compared with Ramrikawn (peri-urban) and Tanhril area (rural area).

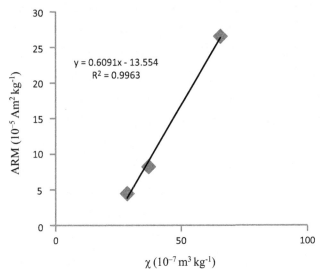

Figure 7.2 Correlation analysis of χ and ARM of *Ficus bengalensis* leaf.

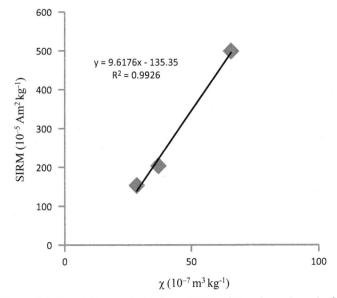

Figure 7.3 Correlation analysis of χ and SIRM of *Ficus bengalensis* leaf.

Significant correlation coefficients have been recorded between all magnetic parameters of *F. bengalensis* plant leaves. It was also observed that all magnetic parameters showed significant correlation with Fe, Pb, and Cu (Figures 7.4–7.12; Table 7.3). The observations indicated that magnetic

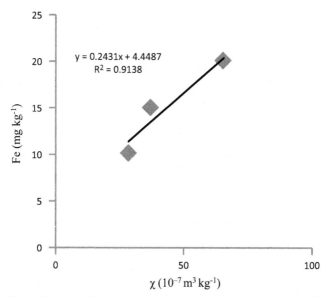

Figure 7.4 Correlation between Fe and χ of *Ficus bengalensis* leaf.

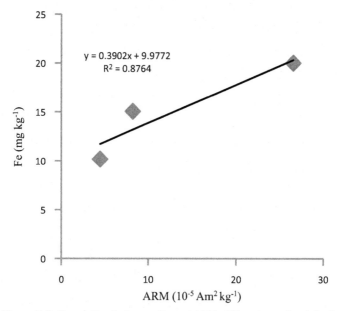

Figure 7.5 Correlation between Fe and ARM of *Ficus bengalensis* leaf.

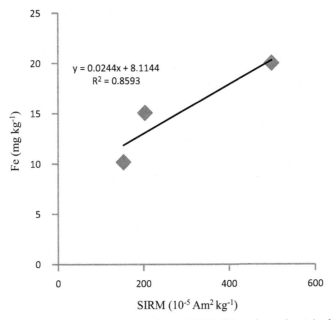

Figure 7.6 Correlation between Fe and SIRM of *Ficus bengalensis* leaf.

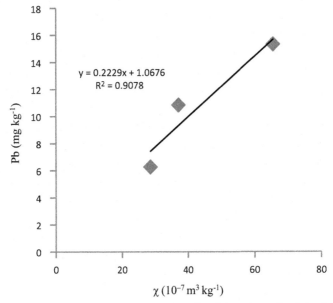

Figure 7.7 Correlation between Pb and χ of *Ficus bengalensis* leaf.

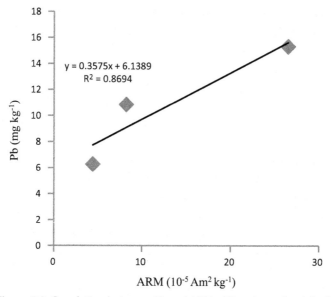

Figure 7.8 Correlation between Pb and ARM of *Ficus bengalensis* leaf.

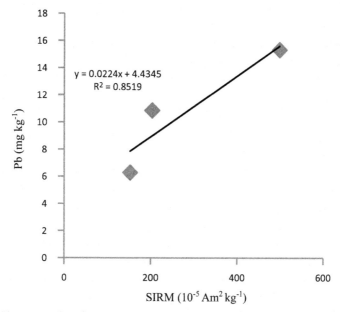

Figure 7.9 Correlation between Pb and SIRM of *Ficus bengalensis* leaf.

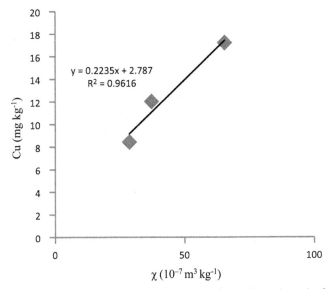

Figure 7.10 Correlation between Cu and χ of *Ficus bengalensis* leaf.

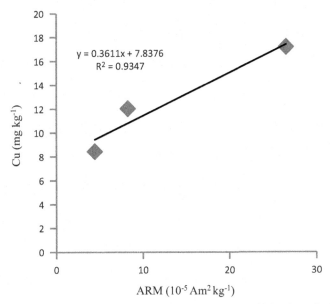

Figure 7.11 Correlation between Cu and ARM of *Ficus bengalensis* leaf.

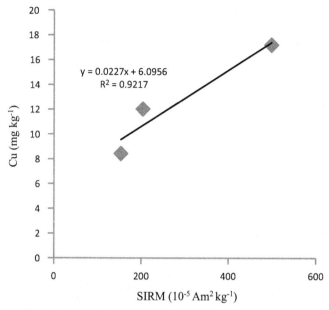

Figure 7.12 Correlation between Cu and SIRM of *Ficus bengalensis* leaf.

Table 7.3 Fe, Pb, and Cu contents (mg kg^{-1}) with standard deviation (S.D.) of *Ficus bengalensis* leaf

Mean contents of Fe, Pb, and Cu (mg kg^{-1}) ± S.D.

Sampling site	Fe content	Pb content	Cu content
Zarkawt	20.01 ± 0.17	15.32 ± 0.51	17.21 ± 0.92
Tanhril	10.17 ± 0.31	6.27 ± 0.18	8.44 ± 0.38
Ramrikawn	15.06 ± 0.11	10.85 ± 0.41	12.03 ± 0.16

properties of dust-loaded particles are rich in ferromagnetic minerals and may act as a proxy for ambient PM pollution levels.

7.4 CONCLUSIONS

According to our results from the study on *F. bengalensis* tree leaves in Aizawl city, we can conclude the following: (1) Magnetic properties of tree leaves change significantly in different functional areas. Overall all values of magnetic parameters (χ, ARM, and SIRM) decline in the following sequence: city area > peri-urban area > rural area. Magnetic concentration data suggest that the deposition of PM on tree leaves varies due to different traffic behavior between sites and due to other activities like soil erosion, mining, and stone quarrying, etc. (2) The magnetic properties of tree leaves in Aizawl city revealed that the magnetic fraction of dust is dominated by multidomain magnetite-like ferromagnetic particles. (3) Magnetic survey of tree leaves is recommended as an inexpensive tool, i.e., tree leaves are easy to collect and measure. Biomagnetic monitoring of PM through plant leaves provides a rapid and economical technique for monitoring atmospheric PM pollution, thus paving the way for innovation of an ecosustainable environmental management tool.

REFERENCES

Champion, H., Seth, S.K., 1968. A Revised Survey of Forest Types of India. Government of India Press, Delhi.

Chauhan, A., Pawar, M., 2010. Assessment of ambient air quality status in urbanization, industrialization and commercial centers of Uttarakhand (India). New York Science Journal 3 (7), 85–94.

Dearing, J., 1999. Magnetic susceptibility. In: Walden, J., Oldfield, F., Smith, J.P. (Eds.), Environmental Magnetism: A Practical Guide. Technical Guide No. 6. Quaternary Research Association, Cambridge, England, pp. 35–62.

Englert, N., 2004. Fine particles and human health—a review of epidemiological studies. Toxicology Letters 149, 235–242.

Evans, M.E., Heller, F., 2003. Environmental Magnetism: Principles and Applications of Enviromagnetics (International Geophysics). Academic Press, London. Elsevier 299 pp.

Faiz, Y., Tufail, M., Javed, M.T., Chaudhry, M.M., Siddique, N., 2009. Road dust pollution of Cd, Cu, Ni, Pb and Zn along Islamabad Expressway, Pakistan. Microchemical Journal 92, 186–192.

Harrison, R.M., Jones, M., 1995. The chemical composition of airborne particles in the UK atmosphere. Science of the Total Environment 168 (3), 195–214.

Huhn, G., Schulz, H., Staerk, H.J., Toelle, R., Scheuermann, G., 1995. Evaluation of regional heavy metal deposition by multivariate analysis of element contents in pine tree barks. Water, Air, and Soil Pollution 84 (3–4), 367–383.

Janssen, N.A.H., Lanki, T., Hoek, G., Vallius, M., de Hartog, J.J., Van Grieken, R., 2005. Associations between ambient, personal and indoor exposure to fine particulate matter constituents in Dutch and Finnish panels of cardiovascular patients. Occupational and Environmental Medicine 62, 868–877.

Jerrett, M., Buzzelli, M., Burnett, R.T., De Luca, P.F., 2005. Particulate air pollution, social confounders, and mortality in small areas of an industrial city. Social Science and Medicine 60, 2845–2863.

King, J.W., Channell, J.E.T., 1991. Sedimentary magnetism, environmental magnetism and magnetostratigraphy. Reviews of Geophysics 29, 358–370.

Knox, E.G., 2006. Roads, railways and childhood cancers. Journal of Epidemiology and Community Health 60 (2), 136–141.

Knutsen, S., Shavlik, D., Chen, L.H., Beeson, W.L., Ghamsary, M., Petersen, F., 2004. The association between ambient particulate air pollution levels and risk of cardiopulmonary and all-cause mortality during 22 years follow-up of a non-smoking cohort. Epidemiology 15 (4), S45.

Laltlanchhuanga, S.K., 2006. Studies of the Impact of Disturbance on Secondary Productivity of Forest Ecosystem with Special Reference to Surface, Sub-surface Litter Insect and Other Non-Insect Groups. Mizoram University (M.Sc. dissertation).

Le Tertre, A., Medina, S., Samoli, E., Forsberg, B., Michelozzi, P., Boumghar, A., 2002. Short-term effects of particulate air pollution on cardiovascular diseases in eight European cities. Journal of Epidemiology and Community Health 56, 773–779.

Maher, B.A., Matzka, J., 1999. Magnetic biomonitoring of roadside tree leaves; identification of spatial and temporal variation in vehicle derived particulates. Atmospheric Environment 33, 4565–4569.

Maher, B.A., Thompson, R., 1999. Quaternary, Climates, Environments and Magnetism. Cambridge University Press. 390 pp.

Maher, B.A., Mitchell, R., Kinnersley, R., 2010. High-resolution magnetic biomonitoring: a quantitative surrogate for particulate pollution. In: CLIMAQS Workshop 'Local Air Quality and its Interactions with Vegetation'.

Morawska, L., Zhang, J., 2002. Combustion sources of particles. 1. Health relevance and source signatures. Chemosphere 49, 1045–1058.

Moreno, E., Sagnotti, L., Dinares-Turell, J., Winkler, A., Cascella, A., 2003. Biomonitoring of traffic air pollution in Rome using magnetic properties of tree leaves. Atmospheric Environment 37, 2967–2977.

Muxworthy, A.R., Matzka, J., Davila, A.F., Petersen, N., 2003. Magnetic signature of daily sampled urban atmospheric particles. Atmospheric Environment 37, 4163–4169.

Pope III, C.A., Burnett, R.T., Thun, M.J., Calle, E.E., Krewski, D., Ito, K., Thurston, G.D., 2002. Lung cancer, cardiopulmonary mortality, and long-term exposure to fine particulate air pollution. Journal of American Medical Association 287 (9), 1132–1141.

Power, A.L., Worsley, A.T., Booth, C., 2009. Magneto-biomonitoring of intra-urban spatial variations of particulate matter using tree leaves. Environmental Geochemistry and Health 31, 315–325.

Rai, P.K., Panda, L.S., Chutia, B.M., Singh, M.M., 2013. Comparative assessment of air pollution tolerance index (APTI) in the industrial (Rourkela) and non industrial area (Aizawl) of India: an eco-management approach. African Journal of Environmental Science and Technology 7 (10), 944–948.

Rai, P.K., Panda, L.S., 2014. Dust capturing potential and air pollution tolerance index (APTI) of some roadside tree vegetation in Aizawl, Mizoram, India: an Indo-Burma hot spot region. Air Quality, Atmosphere and Health 7 (1), 93–101.

Rai, P.K., 2013. Environmental magnetic studies of particulates with special reference to biomagnetic monitoring using roadside plant leaves. Atmospheric Environment 72, 113–129.

Rai, P.K., Chutia, B., 2015. Biomonitoring of atmospheric particulate matter (PM) using magnetic properties of *Ficus bengalensis* tree leaves in Aizawl, Mizoram, North-East India. International Journal of Environmental Sciences 5 (4), 856–869.

Robinson, S.G., 1986. The late Pleistocene palaeoclimatic record of North Atlantic deep-sea sediments revealed by mineral-magnetic measurements. Physics of the Earth and Planetary Interiors 42 (1–2), 22–47.

Schwartz, J., 1996. Air pollution and hospital admissions for respiratory disease. Epidemiology 7 (1), 20–28.

Sun, Z., Hu, S., Ma, X., 1996. A rock-magnetic study of recent lake sediments and its palaeo environmental implication. Acta Geophysica Sinica 39 (2), 178–187.

Szonyi, M., Sagnotti, L., Hirt, A.M., 2008. A refined biomonitoring study of airborne particulate matter pollution in Rome, with magnetic measurements on *Quercus ilex* tree leaves. Geophysical Journal International 173 (1), 127–141.

Thompson, R., 1986. Modelling magnetization data using SIMPLEX. Physics of the Earth and Planetary Interiors 42, 113–127.

Thompson, R., Oldfield, F., 1986. Environmental Magnetism. Allen and Unwin, London. 227 pp.

Urbat, M., Lehndorff, E., Schwark, L., 2004. Biomonitoring of air quality in Cologne conurbation using pine needles as a passive sampler. Part I: magnetic properties. Atmospheric Environment 38, 3781–3792.

Walden, J., 1999. Sample collection and preparation. In: Walden, J., Oldfield, F., Smith, J.P. (Eds.), Environmental Magnetism: A Practical Guide. Technical Guide No. 6. Quaternary Research Association, Cambridge, England, pp. 26–34.

Yin, G., Hu, S., Cao, L., Roesler, W., Appel, E., 2013. Magnetic properties of tree leaves and their significance in atmospheric particle pollution in Linfen city, China. Chinese Geographical Science 23 (1), 59–72.

Yu, L., Xu, Y., Zhang, W., 1995. Magnetic measurement on lake sediment and its environmental application. Progress in Geophysics 10 (1), 11–22.

Zhang, C.X., Huang, B.C., Li, Z.Y., Liu, H., 2006. Magnetic properties of highroad-side pine tree leaves in Beijing and their environmental significance. Chinese Science Bulletin 51 (24), 3041–3052.

CHAPTER EIGHT

Biomonitoring of Particulate Pollution through Magnetic Properties of Important Horticultural Plant Species

8.1 INTRODUCTION

Atmospheric particulate matter (PM) is considered to be the largest contributor to the problem of urban air pollution (Szonyi et al., 2008). There is a strong correlation between PM and respiratory health damage (Schwartz, 1996; Pope et al., 2002; Knutsen et al., 2004; Knox, 2006; Yin et al., 2013). Many studies highlight the importance of particulates with an aerodynamic diameter of less than $10\,\mu m$ (PM_{10}), which, due to their small size, can penetrate deep into the human lung and cause respiratory illness (Le Tertre et al., 2002; Janssen et al., 2005; Jerrett et al., 2005). Alongside PM_{10} are further grain size divisions of $PM_{2.5}$ and $PM_{0.1}$ ($2.5\,\mu m$ and $0.1\,\mu m$, respectively, again relative to their aerodynamic diameters). These fine and ultrafine particulates have higher burdens of toxicity as they become coated with heavy metals and chemicals, which, when inhaled, can become absorbed into the body and may target specific organs (Morawska and Zhang, 2002; Englert, 2004; Power et al., 2009). Urban anthropogenic PM contains certain heavy metals that are toxic to human health (Harrison and Jones, 1995; Huhn et al., 1995). In view of the abovementioned deleterious impacts of particulate matter, it is quite obvious that it is important to investigate feasible and ecosustainable green technologies. Although there are many conventional (physical and chemical) devices for assessment of air pollution, plant systems allow the direct assessment of the air stressors particularly in the context of magnetic particles (Maher et al., 2010). Biological monitors are organisms that provide quantitative information on some aspects of the environment, such as how much of a pollutant is present. In this regard, the air-cleansing capacity of urban trees presents an alternative approach to foster an

Biomagnetic Monitoring of Particulate Matter
ISBN 978-0-12-805135-1
http://dx.doi.org/10.1016/B978-0-12-805135-1.00008-1

155

integrated approach to the sustainable management of urban ecosystems. Moreover, in urban areas higher plants are mostly suitable for monitoring dust pollution as lichens and mosses are often missing (Faiz et al., 2009).

Plants are good indicators of air pollution. Tree leaves have proven to be good collectors of PM (Moreno et al., 2003; Urbat et al., 2004; Yin et al., 2013; Rai, 2013). Vegetation naturally cleanses the atmosphere by absorbing some particulate matter and gases through plant leaves, as they are continuously exposed to the surrounding atmosphere and are therefore the main receptor of particulate pollutants. It has been demonstrated that magnetic measurement is an important means in particulate pollution study through plant leaves. Biomonitoring of particulate pollution through magnetic properties of plant leaves is a reliable, rapid, and inexpensive alternative to conventional atmospheric pollution monitoring (Power et al., 2009). This has promoted its suitability for aiding biomonitoring of air quality (Maher and Matzka, 1999; Moreno et al., 2003; Urbat et al., 2004). The magnetic properties of tree leaves as a proxy in monitoring and mapping of PM pollution have drawn increasing attention (Zhang et al., 2006).

In our research, we carried out a primary magnetic study on PM pollution in Aizawl city, Mizoram. The rapid urbanization, fast, drastic increases in number of vehicles on the roads and other activities including soil erosion, mining, stone quarrying, and shifting cultivation in Aizawl, have led to increases in the concentration of particulate pollution in the atmosphere. In this study, we collected three horticulturally important tree leaves (*Artocarpus heterophyllus*, *Psidium guajava*, and *Mangifera indica*) from different parts of Aizawl city and conducted a series of environmental and particulate magnetic measurements, trying to map the PM pollution, to provide essential data for the recognition and control of air quality as well as for further environmental study.

8.2 DESCRIPTION OF STUDY SITE

Mizoram ($21°56'–24°31'N$ and $92°16'–93°26'E$) is one of the eight states of northeast India (Figure 8.1), covering an area of $21,081 \, km^2$. The Tropic of Cancer divides the state into two almost equal parts. The state is bordered by Myanmar to the east and south, Bangladesh to the west, and the states of Assam, Manipur, and Tripura to the north. The altitude rises towards the Myanmar border. The forest vegetation of state can be divided into three major categories: tropical wet evergreen forest, tropical semi-evergreen forest, and subtropical pine forest (Champion and Seth, 1968).

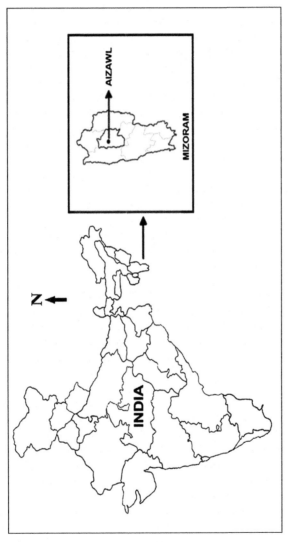

Figure 8.1 Map of the study area.

Aizawl (21°58′–21°85′N and 90°30′– 90°60′E), the capital of the state is 1132 m above sea level (asl). The altitude in Aizawl district varies from 800 to 1200 m asl. The climate of the area is typically monsoonal. The annual average rainfall is c. 2350 mm. The area experiences distinct seasons. The ambient air temperature normally is in the range of 20–30 °C in summer and 11–21 °C in winter (Laltlanchhuanga, 2006).

The study was carried out in Aizawl district, which was categorized into three subsites. *Site 1. Zarkawt:* Zarkawt is a commercial place in the city of Aizawl. Because of high traffic density, the emission of dust particles is seen very high. *Site 2. Ramrikawn:* It is peri-urban and commercial area with markets, bus stands, and food storage (Food Corporation of India). *Site 3. Tanhril:* It is a rural area with low vehicular activity, located in the western part of Aizawl district.

8.3 MATERIALS AND METHODS

Sampling was conducted during the months of February, March, April, and May 2013. Tree leaves were collected from three species on dry sunny days. The recorded plants were *A. heterophyllus*, *P. guajava*, and *M. indica*. These three plants were selected for the study because of their availability and commonness. At each site, five leaves of similar size were collected from the tree on the side nearest to the road at a height of approximately 2 m to avoid possible contamination from ground splash. The leaves were dried at 35 °C and the dried weight was recorded; samples were prepared for magnetic analysis, which involved packing the dried leaves into 10-cc plastic sample pots (Walden, 1999).

8.3.1 Magnetic Parameters

The magnetic parameters such as magnetic susceptibility (χ), anhysteretic remanent magnetization (ARM), and saturation isothermal remanent magnetization (SIRM) were carried out at K.S. Krishnan Geomagnetic Research Lab of the Indian Institute of Geomagnetism, Allahabad, India. The magnetic susceptibility reflects the total composition of the dust deposited on the leaves, with a prevailing contribution from ferromagnetic minerals, which have much higher susceptibility values than paramagnetic and diamagnetic minerals, such as, clay or quartz (Maher and Thompson, 1999; Evans and Heller, 2003). A Bartington (Oxford, England) MS2B dual frequency susceptibility meter was used (Dearing, 1999) and measurements were taken.

Anhysteretic remanent magnetization indicates the magnetic concentration and is also sensitive to the presence of fine grains ~0.04–1 μm (Thompson and Oldfield, 1986), thus, falling within the respirable size range of $PM_{2.5}$ and having the potential to have a high burden of toxicity (Power et al., 2009). The ARM was induced in samples using a Molspin (Newcastle-upon-Tyne, England) AF demagnetizer, whereby a direct current biasing field is generated in the presence of an alternating field, which peaks at 100 milli-Tesla (mT). The nature of this magnetic field magnetizes the fine magnetic grains and the amount of magnetization retained within the sample (remanence) when removed from the field was measured using a Molspin1A magnetometer. The samples were then demagnetized to remove this induced field in preparation for the subsequent magnetic analysis (Walden, 1999).

Saturation isothermal remanent magnetization indicates the total concentration of magnetic grains (Evans and Heller, 2003) and can be used as a proxy of particulate matter concentration (Muxworthy et al., 2003). It involves measuring the magnetic remanence of samples once removed from an induced field. Using a Molspin pulse magnetizer, SIRM of 800 mT in the forward field was induced in the samples. At this high magnetization, all magnetic grains within the sample become magnetized (Power et al., 2009).

The ratio of IRM-300 and SIRM was defined as the S-ratio (King and Channell, 1991). The S-ratio mainly reflects the relative proportion of antiferromagnetic to ferrimagnetic minerals in a sample. A ratio close to 1.0 reflects almost pure magnetite while ratios of <0.8 indicate the presence of some antiferromagnetic minerals, generally goethite or hematite (Thompson, 1986).

8.3.2 Statistical Analysis

Correlation coefficient values were calculated at each site using SPSS software (SPSS Inc., version 10.0).

8.4 RESULTS AND DISCUSSION

The ambient PM concentrations were recorded highest at Zarkawt, followed by Ramrikawn, while lowest values were recorded for the Tanhril area. The average magnetic data collected throughout the four-month sampling period is presented in Tables 8.1–8.3, respectively, for all three tree leaves (*A. heterophyllus, P. guajava,* and *M. indica*).

Table 8.1 Statistics of magnetic properties (mean and standard deviation) of *Artocarpus heterophyllus* leaf

Site	χ (10^{-7} m³ kg⁻¹)	ARM (10^{-5} Am² kg⁻¹)	SIRM (10^{-5} Am² kg⁻¹)	ARM/χ (10^2 Am⁻¹)	SIRM/χ (10^2 Am⁻¹)	S-ratio
Zarkawt	38.21±0.42	23.10±0.31	265.21±0.61	0.60	6.94	0.95
Tanhril	26.81±0.25	4.46±0.23	153.11±0.27	0.16	5.71	0.96
Ramrikawn	34.62±0.29	9.53±0.38	273.41±0.63	0.27	7.89	0.95

Table 8.2 Statistics of magnetic properties (mean and standard deviation) of *Psidium guajava* leaf

Site	χ (10^{-7} m³ kg⁻¹)	ARM (10^{-5} Am² kg⁻¹)	SIRM (10^{-5} Am² kg⁻¹)	ARM/χ (10^2 Am⁻¹)	SIRM/χ (10^2 Am⁻¹)	S-ratio
Zarkawt	37.09±0.81	8.24±0.31	203.70±0.52	0.22	5.49	0.95
Tanhril	28.59±0.39	4.48±0.29	153.21±0.31	0.15	5.35	0.96
Ramrikawn	33.87±0.54	8.19±0.41	201.42±0.26	0.24	5.94	0.95

Table 8.3 Statistics of magnetic properties (mean and standard deviation) of *Mangifera indica* leaf

Site	χ (10^{-7} m³ kg⁻¹)	ARM (10^{-5} Am² kg⁻¹)	SIRM (10^{-5} Am² kg⁻¹)	ARM/χ (10^2 Am⁻¹)	SIRM/χ (10^2 Am⁻¹)	S-ratio
Zarkawt	44.78±0.15	40.74±0.49	292.62±0.77	0.90	6.53	0.95
Tanhril	29.01±0.38	5.09±0.73	153.83±0.34	0.17	5.30	0.96
Ramrikawn	39.19±0.42	12.76±0.29	266.11±0.61	0.32	6.79	0.95

In Zarkawt, the χ, ARM, and SIRM values were 38.21 ± 0.42 ($10^{-7}\,\mathrm{m^3\,kg^{-1}}$), 23.10 ± 0.31 ($10^{-5}\,\mathrm{Am^2\,kg^{-1}}$), and 265.21 ± 0.61 ($10^{-5}\,\mathrm{Am^2\,kg^{-1}}$), respectively for *A. heterophyllus*. For *P. guajava*, the χ value was 37.09 ± 0.81 ($10^{-7}\,\mathrm{m^3\,kg^{-1}}$), ARM value was 8.24 ± 0.31 ($10^{-5}\,\mathrm{Am^2\,kg^{-1}}$), and SIRM value was 203.70 ± 0.52 ($10^{-5}\,\mathrm{Am^2\,kg^{-1}}$). The χ, ARM, and SIRM values were 44.78 ± 0.15 ($10^{-7}\,\mathrm{m^3\,kg^{-1}}$), 40.74 ± 0.49 ($10^{-5}\,\mathrm{Am^2\,kg^{-1}}$), and 292.62 ± 0.77 ($10^{-5}\,\mathrm{Am^2\,kg^{-1}}$), respectively, for *M. indica*.

In Tanhril, it was found that the χ value of *A. heterophyllus* was 26.81 ± 0.25 ($10^{-7}\,\mathrm{m^3\,kg^{-1}}$), ARM was 4.46 ± 0.23 ($10^{-5}\,\mathrm{Am^2\,kg^{-1}}$), and SIRM was 153.11 ± 0.27 ($10^{-5}\,\mathrm{Am^2\,kg^{-1}}$). Similarly, *P. guajava* had a value of 28.59 ± 0.39 ($10^{-7}\,\mathrm{m^3\,kg^{-1}}$) for χ, 4.48 ± 0.29 ($10^{-5}\,\mathrm{Am^2\,kg^{-1}}$) for ARM, and 153.21 ± 0.31($10^{-5}\,\mathrm{Am^2\,kg^{-1}}$) for SIRM. The χ, ARM, and SIRM values were 29.01 ± 0.38 ($10^{-7}\,\mathrm{m^3\,kg^{-1}}$), 5.09 ± 0.73 ($10^{-5}\,\mathrm{Am^2\,kg^{-1}}$), and 153.83 ± 0.34 ($10^{-5}\,\mathrm{Am^2\,kg^{-1}}$), respectively, for *M. indica*.

In Ramrikawn, the χ, ARM, and SIRM values were 34.62 ± 0.29 ($10^{-7}\,\mathrm{m^3\,kg^{-1}}$), 9.53 ± 0.38 ($10^{-5}\,\mathrm{Am^2\,kg^{-1}}$), and 273.41 ± 0.63 ($10^{-5}\,\mathrm{Am^2\,kg^{-1}}$), respectively, for *A. heterophyllus*. For *P. guajava*, χ value was 33.87 ± 0.54 ($10^{-7}\,\mathrm{m^3\,kg^{-1}}$), ARM value was 8.19 ± 0.41 ($10^{-5}\,\mathrm{Am^2\,kg^{-1}}$), and SIRM value was 201.42 ± 0.26 ($10^{-5}\,\mathrm{Am^2\,kg^{-1}}$). The χ, ARM, and SIRM values were 39.19 ± 0.42 ($10^{-7}\,\mathrm{m^3\,kg^{-1}}$), 12.76 ± 0.29 ($10^{-5}\,\mathrm{Am^2\,kg^{-1}}$), and 266.11 ± 0.61 ($10^{-5}\,\mathrm{Am^2\,kg^{-1}}$), respectively, for *M. indica*.

The high dispersion degrees of susceptibility and remanent magnetism mainly result from the sampling sites in different functional areas. Samples collected in the rural area show low susceptibility and remanent magnetism, whereas tree leaves sampled in city and peri-urban areas show higher values. The correlation of magnetic susceptibility with ARM and SIRM are significant (Figures 8.2–8.7). The relatively high correlation indicates that the magnetic minerals with paramagnetism and superparamagnetism contribute slightly to the magnetism of tree leaves, and the major contributor is ferro(i)magnetic minerals (Yu et al., 1995: Sun et al., 1996). The values of ARM/χ and SIRM/χ can reflect the grain size of magnetic minerals (Thompson and Oldfield, 1986; Evans and Heler, 2003). From the study it was observed that ARM/χ and SIRM/χ values are low at all study sites (Tables 8.1–8.3). Low values of ARM/χ and SIRM/χ indicate relatively large grain size magnetic particles in leaf samples (Yin et al., 2013). S-ratio of all three leaf samples ranges from 0.95 to 0.96 (Tables 8.1–8.3), which means that these leaf samples are dominated by "soft" magnetic minerals with a low coercive force, but a minor part of "hard" magnetic minerals with a relatively high coercive force also exists (Robinson, 1986).

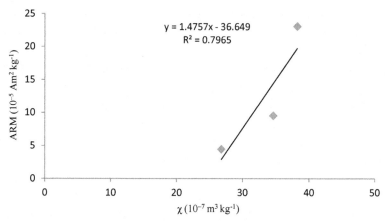

Figure 8.2 Correlation analysis of magnetic susceptibility (χ) and anhysteretic remanent magnetization (ARM) of *Artocarpus heterophyllus* leaf.

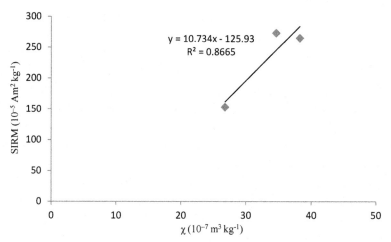

Figure 8.3 Correlation analysis of magnetic susceptibility (χ) and saturation isothermal remanent magnetization (SIRM) of *Artocarpus heterophyllus* leaf.

From the findings recorded in Tables 8.1–8.3 we can infer that the magnetic values for all three species display similar trends, with Zarkawt representing the highest and Tanhril area representing the lowest concentration data. Further, results indicate that Zarkawt and Ramrikawn experience the highest deposition of magnetic grains originating from PM. The χ, ARM, and SIRM values were high for *M. indica* when compared with *A. heterophyllus* and *P. guajava*.

The average magnetic concentration data (Tables 8.1–8.3) demonstrate that the accumulation of PM on tree leaves varies across the three locations.

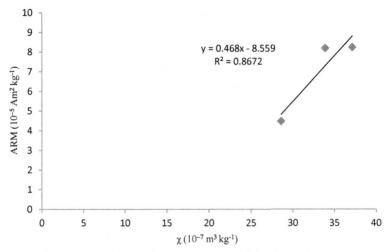

Figure 8.4 Correlation analysis of magnetic susceptibility (χ) and anhysteretic remanent magnetization (ARM) of *Psidium guajava* leaf.

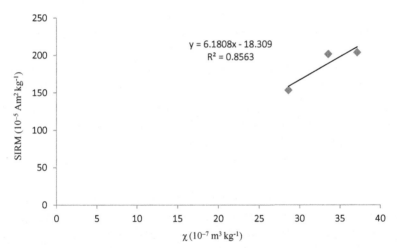

Figure 8.5 Correlation analysis of magnetic susceptibility (χ) and saturation isothermal remanent magnetization (SIRM) of *Psidium guajava* leaf.

The results suggest that Zarkawt and Ramrikawn experience the heaviest loads of particulates in comparison to the low-depositions site Tanhril area. This suggests that localized conditions like environmental, metrological, or anthropogenic may be influencing or disturbing particulate deposition or it may reflect differences in the ability of leaf species to capture particulates (Power et al., 2009). Zarkawt recorded the highest values of magnetic parameters, which may be attributed to heavy vehicle load (due to city area) compared with Ramrikawn (peri-urban) and Tanhril area (rural area).

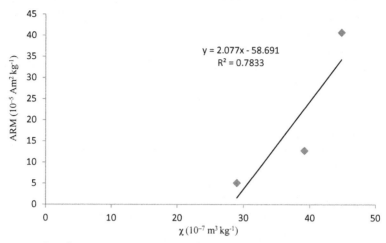

Figure 8.6 Correlation analysis of magnetic susceptibility (χ) and anhysteretic remanent magnetization (ARM) of *Mangifera indica* leaf.

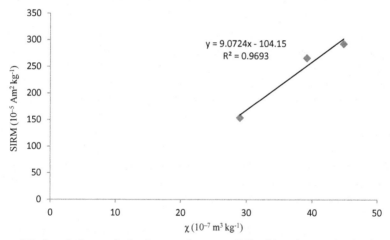

Figure 8.7 Correlation analysis of magnetic susceptibility (χ) and saturation isothermal remanent magnetization (SIRM) of *Mangifera indica* leaf.

8.5 CONCLUSIONS

Biomonitoring of atmospheric particulate matter using magnetic properties of tree leaves is a useful approach to delineate primary anthropogenic airborne particulate pollution, which leads to the deterioration of ambient air quality and causes adverse effects to human health. According to our preliminary results from the study on tree leaves in Aizawl city, we can conclude the following: (1) Magnetic properties of tree leaves change significantly in different functional areas. Overall, all values of magnetic parameters (χ, ARM, and SIRM) decline in the following sequence: city area > peri-urban area > rural area. Magnetic concentration data suggest that the deposition of PM on tree leaves varies due to different traffic behavior between sites and due to other activities like soil erosion, mining, stone quarrying, etc. (2) The magnetic properties of tree leaves in Aizawl city revealed that the magnetic fraction of dust is dominated by multidomain magnetite-like ferromagnetic particles. (3) Magnetic survey of tree leaves is recommended as an inexpensive tool, i.e., tree leaves are easy to collect and measure. The magnetic analysis of dust loadings on tree leaves provides an alternative proxy method to conventional air pollution monitoring.

REFERENCES

Champion, H., Seth, S.K., 1968. A Revised Survey of Forest Types of India. Government of India Press, Delhi.

Dearing, J., 1999. Magnetic susceptibility. In: Walden, J., Oldfield, F., Smith, J.P. (Eds.), Environmental Magnetism: A Practical Guide. Technical Guide No. 6. Quaternary Research Association, Cambridge, England, pp. 35–62.

Englert, N., 2004. Fine particles and human health—a review of epidemiological studies. Toxicology Letters 149, 235–242.

Evans, M.E., Heller, F., 2003. Environmental Magnetism: Principles and Applications of Enviromagnetics (International Geophysics). Academic Press. Elsevier, London. 299 pp.

Faiz, Y., Tufail, M., Javed, M.T., Chaudhry, M.M., Siddique, N., 2009. Road dust pollution of Cd, Cu, Ni, Pb and Zn along Islamabad Expressway, Pakistan. Microchemical Journal 92, 186–192.

Harrison, R.M., Jones, M., 1995. The chemical composition of airborne particles in the UK atmosphere. Science of the Total Environment 168 (3), 195–214.

Huhn, G., Schulz, H., Staerk, H.J., Toelle, R., Scheuermann, G., 1995. Evaluation of regional heavy metal deposition by multivariate analysis of element contents in pine tree barks. Water, Air, and Soil Pollution 84 (3–4), 367–383.

Janssen, N.A.H., Lanki, T., Hoek, G., Vallius, M., de Hartog, J.J., Van Grieken, R., 2005. Associations between ambient, personal and indoor exposure to fine particulate matter constituents in Dutch and Finnish panels of cardiovascular patients. Occupational and Environmental Medicine 62, 868–877.

Jerrett, M., Buzzelli, M., Burnett, R.T., DeLuca, P.F., 2005. Particulate air pollution, social confounders, and mortality in small areas of an industrial city. Social Science and Medicine 60, 2845–2863.

King, J.W., Channell, J.E.T., 1991. Sedimentary magnetism, environmental magnetism and magnetostratigraphy. Reviews of Geophysics 29, 358–370.

Knox, E.G., 2006. Roads, railways and childhood cancers. Journal of Epidemiology and Community Health 60 (2), 136–141.

Knutsen, S., Shavlik, D., Chen, L.H., Beeson, W.L., Ghamsary, M., Petersen, F., 2004. The association between ambient particulate air pollution levels and risk of cardiopulmonary and all-cause mortality during 22 years follow-up of a non-smoking cohort. Epidemiology 15 (4), S45.

Laltlanchhuanga, S.K., 2006. Studies of the Impact of Disturbance on Secondary Productivity of Forest Ecosystem with Special Reference to Surface, Sub-surface Litter Insect and Other Non-Insect Groups. Mizoram University (M.Sc. dissertation).

Le Tertre, A., Medina, S., Samoli, E., Forsberg, B., Michelozzi, P., Boumghar, A., 2002. Short-term effects of particulate air pollution on cardiovascular diseases in eight European cities. Journal of Epidemiology and Community Health 56, 773–779.

Maher, B.A., Matzka, J., 1999. Magnetic biomonitoring of roadside tree leaves; identification of spatial and temporal variation in vehicle derived particulates. Atmospheric Environment 33, 4565–4569.

Maher, B.A., Mitchell, R., Kinnersley, R., 2010. High-resolution magnetic biomonitoring: a quantitative surrogate for particulate pollution. In: CLIMAQS Workshop 'Local Air Quality and its Interactions with Vegetation'.

Maher, B.A., Thompson, R., 1999. Quaternary, Climates, Environments and Magnetism. Cambride University Press. 390 pp.

Morawska, L., Zhang, J., 2002. Combustion sources of particles. 1. Health relevance and source signatures. Chemosphere 49, 1045–1058.

Moreno, E., Sagnotti, L., Dinares-Turell, J., Winkler, A., Cascella, A., 2003. Biomonitoring of traffic air pollution in Rome using magnetic properties of tree leaves. Atmospheric Environment 37, 2967–2977.

Muxworthy, A.R., Matzka, J., Davila, A.F., Petersen, N., 2003. Magnetic signature of daily sampled urban atmospheric particles. Atmospheric Environment 37, 4163–4169.

Pope III, C.A., Burnett, R.T., Thun, M.J., Calle, E.E., Krewski, D., Ito, K., Thurston, G.D., 2002. Lung cancer, cardiopulmonary mortality, and long-term exposure to fine particulate air pollution. Journal of American Medical Association 287 (9), 1132–1141.

Power, A.L., Worsley, A.T., Booth, C., 2009. Magneto-biomonitoring of intra-urban spatial variations of particulate matter using tree leaves. Environmental Geochemistry and Health 31, 315–325.

Rai, P.K., 2013. Environmental magnetic studies of particulates with special reference to biomagnetic monitoring using roadside plant leaves. Atmospheric Environment 72, 113–129.

Robinson, S.G., 1986. The late Pleistocene palaeoclimatic record of North Atlantic deep-sea sediments revealed by mineral-magnetic measurements. Physics of the Earth and Planetary Interiors 42 (1–2), 22–47.

Schwartz, J., 1996. Air pollution and hospital admissions for respiratory disease. Epidemiology 7 (1), 20–28.

Sun, Z., Hu, S., Ma, X., 1996. A rock-magnetic study of recent lake sediments and its palaeoenvironmental implication. Acta Geophysica Sinica 39 (2), 178–187.

Szonyi, M., Sagnotti, L., Hirt, A.M., 2008. A refined biomonitoring study of airborne particulate matter pollution in Rome, with magnetic measurements on *Quercus ilex* tree leaves. Geophysical Journal International 173 (1), 127–141.

Thompson, R., 1986. Modelling magnetization data using SIMPLEX. Physics of the Earth and Planetary Interiors 42, 113–127.

Thompson, R., Oldfield, F., 1986. Environmental Magnetism. Allen and Unwin, London. 227 pp.

Urbat, M., Lehndorff, E., Schwark, L., 2004. Biomonitoring of air quality in Cologne conurbation using pine needles as a passive sampler. Part I: magnetic properties. Atmospheric Environment 38, 3781–3792.

Walden, J., 1999. Sample collection and preparation. In: Walden, J., Oldfield, F., Smith, J.P. (Eds.), Environmental Magnetism: A Practical Guide. Technical Guide No. 6. Quaternary Research Association, Cambridge, England, pp. 26–34.

Yin, G., Hu, S., Cao, L., Roesler, W., Appel, E., 2013. Magnetic properties of tree leaves and their significance in atmospheric particle pollution in Linfen city, China. Chinese Geographical Science 23 (1), 59–72.

Yu, L., Xu, Y., Zhang, W., 1995. Magnetic measurement on lake sediment and its environmental application. Progress in Geophysics 10 (1), 11–22.

Zhang, C.X., Huang, B.C., Li, Z.Y., Liu, H., 2006. Magnetic properties of highroad-side pine tree leaves in Beijing and their environmental significance. Chinese Science Bulletin 51 (24), 3041–3052.

CHAPTER NINE

Biomonitoring of Particulate Matter Using Magnetic Properties (Two-Dimensional Magnetization) of Economically Important Tropical Plant Species

9.1 INTRODUCTION

Atmospheric particulate matter (PM) is one of the most problematic air pollutants in view of its adverse impacts on human health. Atmospheric pollutants exist in both gaseous and particulate form (Bucko et al., 2011). Many studies have highlighted the importance of PM with an aerodynamic diameter of less than $10\,\mu m$ (PM_{10}), which, due to their small size, can penetrate deep into the human lung and cause cardiovascular diseases (Le Tertre et al., 2002; Janssen et al., 2005; Jerrett et al., 2005; Rai, 2011a,b, 2013; Rai et al., 2013; Rai and Panda, 2014). Alongside PM_{10} are further grain size divisions of $PM_{2.5}$ and $PM_{0.1}$ ($2.5\,\mu m$ and $0.1\,\mu m$, respectively, again relative to their aerodynamic diameters). These fine and ultrafine particulates have higher burdens of toxicity as they become coated with heavy metals and chemicals, which, when inhaled, may be absorbed into the body and therefore target specific organs (Morawska and Zhang, 2002; Englert, 2004; Rai, 2013). The urban population is mainly exposed to high levels of air pollution including metals because of motor vehicle emissions, which is also the main source of fine and ultrafine particles (Sharma et al., 2006; Rai, 2013; Rai and Panda, 2014).

In view of the above mentioned deleterious impacts of particulate matter, it is quite obvious that there is a need to investigate feasible and ecosustainable green technologies. Although there are many conventional (physical and chemical) devices for assessment of air pollution, biomonitoring is an efficient tool in urban areas (Rai, 2013). An assessment of particulate pollution through magnetic properties of plant leaves is a rapid and inexpensive alternative to conventional atmospheric pollution monitoring.

Biomagnetic Monitoring of Particulate Matter
ISBN 978-0-12-805135-1
http://dx.doi.org/10.1016/B978-0-12-805135-1.00009-3

Biological monitors are organisms that provide quantitative information on some aspects of the environment, including the presence of pollutants. In this regard, the air cleansing capacity of urban trees presents an alternative means to foster an integrated approach to the sustainable management of urban ecosystems (Rai, 2013). Lichens, bryophytes, or mosses and certain conifers have recently been proven to be potent biomonitoring tools of air pollution (Rai, 2013). However, in urban and peri-urban regions, higher plants are mostly suitable for monitoring dust or PM pollution as lichens and mosses are often absent (Faiz et al., 2009; Rai, 2013). Further, urban trees and shrubs planted in street canyons have proven to be efficient dust-capturing tools (Moreno et al., 2003; Urbat et al., 2004; Rai and Panda, 2014). Spread widely in urban areas and easily collected, tree leaves could improve the scanning resolution in the spatial scale (Mitchell et al., 2010; Gang et al., 2013). With the quick, economical, sensitive, and nondestructive feature of environmental magnetism measurement, the magnetic properties of tree leaves as proxy in monitoring and mapping of PM pollution have drawn increasing attention (Gang et al., 2013). Moreover, tree leaves are efficient passive pollution collectors, as they provide a large surface for particle deposition, a large number of samples and sampling sites, and require no protection from vandalism (Sant'Ovaia et al., 2012). Therefore, urban angiosperm trees offer positive biological, ecological, and aerodynamic effects in comparison to the lower group of plants (Moreno et al., 2003; Urbat et al., 2004; Rai, 2013; Rai et al., 2013). Plants are good indicators of air pollution. Tree leaves have proven to be good collectors of PM (Moreno et al., 2003; Urbat et al., 2004; Yin et al., 2013; Rai, 2013). It has been demonstrated that magnetic measurement is an important means in particulate pollution study through plant leaves. Biomonitoring of particulate pollution through magnetic properties of tree leaves is a reliable, rapid, and inexpensive alternative to conventional atmospheric pollution monitoring (Walden, 1999; Power et al., 2009). This has promoted its suitability for aiding biomonitoring of air quality (Matzka and Maher, 1999; Moreno et al., 2003; Urbat et al., 2004).

The biomagnetic monitoring in urban areas using roadside tree leaves is a new thrust area in the field of PM pollution science. The concept of environmental magnetism as a proxy for atmospheric pollution levels has been reported by several researchers based on analysis of soils and street or roof dust (Hay et al., 1997; Maher, 1998; Hoffmann et al., 1999; Shu et al., 2000; Xie et al., 2000, 2001; Jordanova et al., 2003; Hanesch et al., 2007), and vegetation samples including tree bark (Kletetschka et al., 2003; Urbat et al., 2004). However, increased research has emphasized the use of plant leaves in monitoring the dust or PM (Matzka and Maher, 1999; Moreno et al., 2003; Jordanova et al., 2003; Urbat et al., 2004; Pandey et al., 2005; Maher et al., 2008; Maher, 2009; Kardel et al., 2011; Rai, 2013). Maher and her group were the leaders who performed a cascade of magnetic studies in relation to the environment, thus helping it become a specialized discipline, i.e., environmental geomagnetism (Matzka and Maher, 1999; Maher et al., 2008; Kardel et al., 2011). In view of this, magnetic biomonitoring studies of roadside plant leaves were performed in the Singrauli and Varanasi region of India (Pandey et al., 2005; Prajapati et al., 2006; Sharma et al., 2007; Rai, 2013), in some cities of Portugal, and hilly areas of Nepal (Gautam et al., 2004, 2005), in addition to pioneering work in European countries by several groups led by professor B.A. Maher (e.g., Matzka and Maher, 1999; Maher, 2009; Hansard et al., 2011; Kardel et al., 2011).

However, these studies were confined to temperate plants and only a few studies have been performed on tropical plants pertaining to biomagnetic monitoring (Rai, 2013). In our research, we carried out a primary magnetic study (two-dimensional (2D)-magnetization) on PM pollution in Aizawl city, Mizoram. The rapid urbanization, fast and drastic increases in numbers of vehicles on the roads, and other activities including soil erosion, mining, stone quarrying, and shifting cultivation in Aizawl, have led to increases in the concentration of particulate pollution in the atmosphere. The present study aims to investigate the magnetic properties of different roadside plant leaves at two spatially distant sites in order to compare their capability to accumulate particulates and then mapping the PM pollution, to provide essential data for the recognition and control of air quality as well as for further environmental study.

9.2 MATERIALS AND METHODS
9.2.1 Description of Study Site

Mizoram (21°56′–24°31′N and 92°16′–93°26′E) is one of the eight states under northeast India (Figure 9.1), covering an area of 21,081 km². The Tropic of Cancer divides the state into two almost equal parts. The state borders Myanmar to the east and south, Bangladesh to the west, and the states of Assam, Manipur, and Tripura to the north. The altitude rises towards the Myanmar border. The forest vegetation of state can be classified into three major categories: tropical wet evergreen forest, tropical semi-evergreen forest, and subtropical pine forest (Champion and Seth, 1968).

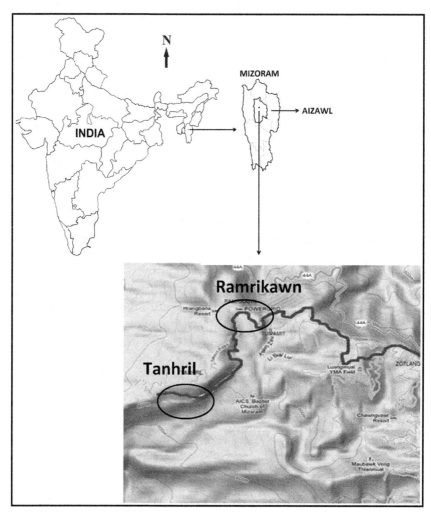

Figure 9.1 Map of the study area, Aizawl, Mizoram, northeast India.

Aizawl (21°58′–21°85′N and 90°30′–90°60′E), the capital of the state, is 1132 m above sea level (asl). The altitude in Aizawl district varies from 800 to 1200 masl. The climate of the area is typically monsoonal, with an annual average rainfall of ca. 2350 mm. The area experiences distinct seasons. The ambient air temperature is normally in the ranges of 20–30 °C in summer and 11–21 °C in winter (Laltlanchhuanga, 2006). It is well known that meteorological data may also affect the air pollutants, including dust or particulate deposition, therefore average meteorological data of the study area recorded during the study period are

Table 9.1 Meteorological data of the study area, i.e., Aizawl, Mizoram

Study period	Temperature		Rainfall (mm)	Humidity (%)
	Maximum (°C)	Minimum (°C)		
September 2013	27.88	20.12	10.12	90.82
October 2013	27.32	19.54	11.07	82.67
November 2013	26.41	15.18	0	69.71
December 2013	24.22	13.28	0	67.39
Average	**26.45**	**17.03**	**5.29**	**77.64**

Border Roads Task Force, Puspak, Aizawl.

mentioned in Table 9.1. Aizawl district falls in the Indo–Burma hotspot region of North East India (Rai, 2009, 2012), and highly diverse plant species having varying leaf morphology can be sampled for dust deposition and study of magnetic parameters. There is great diversity of tropical evergreen plants along the roadsides of Aizawl district, and therefore, these plants can retain the pollutants throughout the year, offering no seasonal constraints to sampling them.

For this present study, Aizawl district was divided into two sites. *Site 1*, Ramrikawn (peri-urban area), is densely populated and the main commercial area of Aizawl district has markets, bus and taxi stands, and the Food Corporation of India (FCI). FCI provides space for food storage for the entire Mizoram state. Due to the presence of FCI in the Ramrikawn area, there is a frequent movement of heavy duty vehicles coming from all parts of India through the national highway of Pushpak (NH-54). As there is are public bus and taxi stands, vehicular movement is usually high in the Ramrikawn area. Stone quarrying activity is also found in this area, also leading to emission of dust particles. Biomass burning through shifting cultivation is also very common in this region (Rai, 2009, 2012) and may be a source of suspended particulate matter pollution. Therefore, in view of these pollution sources, we selected Ramrikawn as a polluted area for investigation. *Site 2*, Tanhril (rural area), with low vehicular activity, is located in the western part of Aizawl district. The load of vehicles is very low and less frequent in comparison to the peri-urban site, and was included as a reference or control site to the peri-urban area.

9.2.2 Sampling

Eight socioeconomically important and evergreen plant species of common occurrence along the roadside were selected for the study: *Mangifera indica, Bougainvillea spectabilis, Artocarpus heterophyllus, Psidium guajava, Hibiscus rosa-sinensis, Cassia fistula, Bauhinia variegata,* and *Delonix regia.* The study was conducted from September to December 2013 for dust deposition and magnetic studies on leaves of selected species. These plants were selected for the study because of their abundance and convenience for sampling. Moreover, these plants have already been investigated for their suitability in efficient dust capturing (Rai et al., 2013; Rai and Panda, 2014). Three replicates of fully matured leaves of each species were randomly collected in early morning from the lower branches (at a height of 2–4 m) and were quickly transferred to the laboratory in polythene bags for further analysis within 24 h of their harvesting.

9.2.3 Dust Deposition

The amount of dust was calculated by taking the initial and final weight of the beaker in which the leaf samples were washed. The following formula was used:

$$W = \left(w_2 - w_1 \right) / a$$

where, W is dust content ($mg\,cm^{-2}$), w_1 is weight of beaker without dust, w_2 is weight of beaker with dust, and a is total area of the leaf (cm^2).

9.2.4 Magnetic Analysis

Mangifera indica, A. heterophyllus, P. Guajava, H. rosa-sinensis, and *B. spectabilis* were selected for magnetic studies due to their high dust deposition capacity among the different plants investigated (Figure 9.2). The magnetic measurements were carried out at K.S. Krishnan Geomagnetic Research Lab of the Indian Institute of Geomagnetism, Allahabad, India. All samples were magnetized with a pulsed magnetic field of 300 milli-Tesla (mT); the isothermal remanent magnetization ($IRM_{300\,mT}$) was then measured with a CCL cryogenic magnetometer with a sensitivity of $10^{-10} Am^2$ (the weakest leaf samples had magnetization of ~$10^{-8} Am^2$). The area of each leaf (cm^2) was calculated by using graph paper. The 2D magnetization was calculated as the magnetic moment per leaf area, in units of amperes ($A = Am^2 m^{-2}$). After measurement, a small number of leaves, representative of different sampling locations, were cleaned with water, detergent, and ultrasonics, to determine their background magnetization.

9.3 RESULTS AND DISCUSSION

Dust deposition capacity of selected plants of two study sites is depicted in Figure 9.2. Data reveal that all plants showed higher dust deposition in

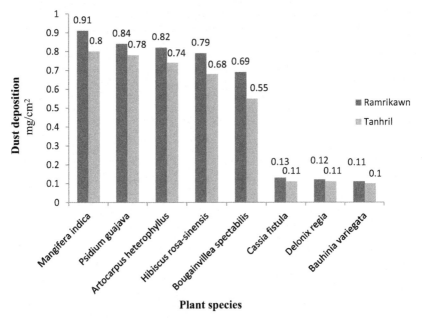

Figure 9.2 Dust deposition capacity (average value) of selected plant species at peri-urban (Ramrikawn) and rural (Tanhril) areas of Aizawl.

the peri–urban (Ramrikawn) area compared to the rural (Tanhril) area. The dust–capturing potential varied from a maximum of $0.91\,\mathrm{mg\,cm^{-2}}$ in *M. indica* to a minimum of $0.1\,\mathrm{mg\,cm^{-2}}$ in *B. variegata*. The trend of dust accumulation among the species was: *M. indica* > *P. guajava* > *A. heterophyllus* > *H. rosa-sinensis* > *B. spectabilis* > *C. fistula* > *D. regia* > *B. variegata* (Figure 9.2). It is observed that the dust deposition capacity depends upon various factors like leaf shape and size, orientation, texture, presence/absence of hairs, length of petioles, etc. The best collector of dust was *M. indica* followed by *P. guajava* and *A. heterophyllus*, which may be attributed to their rough surface, short petiole, and depression in the middle of the leaf. The least dust deposition capacity was found in *B. variegata*, *C. fistula*, and *D. regia* and may be due to smooth surface and no depression in the middle of the leaf samples. Trees with high dust-collecting potential can solve the problems of air particulate pollution to a great extent (Chaphekar, 1972; Dwivedi and Tripathi, 2007; Rai, 2013).

The top five dust-accumulating plants were further tested for their magnetic properties. The 2D magnetization of leaves of five trees (*M. indica*, *P. guajava*, *A. heterophyllus*, *H. rosa-sinensis*, and *B. spectabilis*) at two sites and their background values are shown in Tables 9.2 and 9.3. The higher

Table 9.2 Two-dimensional magnetization (IRM$_{300\,mT}$) of different tree leaves for peri-urban (Ramrikawn) area and background IRM$_{300\,mT}$ of the subsequently cleaned leaves

Plants	2D Magnetization (10^{-6} A)	2D magnetization after cleaning of the leaves (10^{-6} A)	Magnetization removed by cleaning
Mangifera indica	74.14	16.67	86.12
Bougainvillea spectabilis	35.98	7.11	82.72
Artocarpus heterophyllus	52.39	11.93	77.48
Psidium guajava	64.24	14.71	77.72
Hibiscus rosa-sinensis	58.11	13.96	76.29

Table 9.3 Two-dimensional magnetization (IRM$_{300\,mT}$) of different tree leaves for rural (Tanhril) area and background IRM$_{300\,mT}$ of the subsequently cleaned leaves

Plants	2D Magnetization (10^{-6} A)	2D Magnetization after cleaning of the leaves (10^{-6} A)	Magnetization removed by cleaning
Mangifera indica	13.19	2.01	84.28
Bougainvillea spectabilis	8.15	2.84	65.12
Artocarpus heterophyllus	9.88	3.91	63.98
Psidium guajava	11.42	2.14	78.09
Hibiscus rosa-sinensis	10.32	2.41	78.01

magnetization values were recorded the highest at the peri-urban (Ramrikawn) area compared to the rural (Tanhril) area. *Mangifera indica* leaves showed the highest 2D-magnetization value (74.14 × 10^{-6}A) as well as percentage of magnetization removed through cleaning (86.12%) followed by *P. guajava*, *A. heterophyllus*, *H. rosa-sinensis*, and *B. spectabilis* (Tables 9.2 and 9.3) for both study sites.

The magnetic data pertaining to 2D magnetization (Tables 9.2 and 9.3) demonstrate that the accumulation of PM on tree leaves varies across the two locations. The results suggest that the peri-urban (Ramrikawn) area experiences the heaviest loads of particulates in comparison to the low-depositions site, i.e., rural area (Tanhril). This suggests that localized conditions like environmental, metrological or anthropogenic may be influencing or disturbing particulate deposition or it may reflect differences in the ability of leaf species to capture particulates (Power et al., 2009). The peri-urban (Ramrikawn) area recorded the highest values of magnetic parameters specifically in relation to 2D magnetization, which may be attributed to heavy vehicle load (due to location of Food Corporation of India), shifting cultivation, stone mining, street dust, and dust from fragile rocks. Since the rocks of Aizawl are very fragile, the weathered rock dust may also get deposited on plant leaves. The magnetic values were found to be lower in the Tanhril area because of less vehicular pollution and less anthropogenic activates. Biomonitoring of particulate pollution through magnetic properties of tree leaves is recommended as an inexpensive and rapid tool for surveying air pollution in urban areas.

9.4 CONCLUSIONS

Biomagnetic monitoring is the latest approach in the field of PM pollution science. With the advent of environmental magnetism, magnetic measurement is becoming an important means in particulate pollution study. The highest leaf magnetizations were found in the Ramrikawn area, indicating a combustion- and/or exhaust-related source of the magnetic particles. The magnetic analysis of dust loadings on roadside tree leaves provides an alternative proxy method to conventional pollution monitoring. Biomagnetic monitoring of PM through plant leaves provides a rapid and economical technique for monitoring atmospheric PM pollution, thus paving the way for developing an ecosustainable environmental management tool.

REFERENCES

Bucko, M.S., Magiera, T., Johanson, B., Petrovsky, E., Pesonen, L.J., 2011. Identification of magnetic particulates in road dust accumulated on roadside snow using magnetic, geochemical and micro-morphological analysis. Environmental Pollution 159, 1266–1276.

Champion, H., Seth, S.K., 1968. A Revised Survey of Forest Types of India. Government of India Press, Delhi.

Chaphekar, S.B., 1972. Effects of atmospheric pollutants on plants in Bombay. Journal of Biological Sciences 15, 1–6.

Dwivedi, A.K., Tripathi, B.D., 2007. Pollution tolerance and distribution pattern of plants in surrounding area of coal fired industries. Journal of Environmental Biology 28 (2), 257–263.

Englert, N., 2004. Fine particles and human health—a review of epidemiological studies. Toxicology Letters 149, 235–242.

Faiz, Y., Tufail, M., Javed, M.T., Chaudhry, M.M., Siddique, N., 2009. Road dust pollution of Cd, Cu, Ni, Pb and Zn along Islamabad Expressway, Pakistan. Microchemical Journal 92, 186–192.

Gang, Y., Hu, S., Liwan, C., Wolfgang, R., Erwin, A., 2013. Magnetic properties of tree leaves and their significance in atmospheric particle pollution in Linfen City, China. Chinese Geographical Science 23 (1), 59–72.

Gautam, P., Blaha, U., Appel, E., 2005. Magnetic susceptibility of dust-loaded leaves as a proxy of traffic-related heavy metal pollution in Kathmandu city, Nepal. Atmospheric Environment 39, 2201–2211.

Gautam, P., Blaha, U., Appel, E., Neupane, G., 2004. Environmental magnetic approach towards the quantification of pollution in Kathmandu urban area, Nepal. Physics and Chemistry of the Earth, Parts A/B/C 29 (13–14), 973–984.

Hanesch, M., Rantitsch, G., Hemetsberger, S., Scholger, R., 2007. Lithological and pedological influences on the magnetic susceptibility of soil: their consideration in magnetic pollution mapping. Geophysical Journal International 382 (23), 351–363.

Hansard, R., Maher, B.A., Kinnersley, R., 2011. Biomagnetic monitoring of industry-derived particulate pollution. Environmental Pollution 159 (6), 1673–1681.

Hay, K.L., Dearing, J.A., Baban, S.M.J., Loveland, P., 1997. A preliminary attempt to identify atmospherically derived pollution particles in English topsoils from magnetic susceptibility measurements. Physics and Chemistry of the Earth 22, 207–210.

Hoffmann, V., Knab, M., Appel, E., 1999. Magnetic susceptibility mapping of roadside pollution. Journal of Geochemical Exploration 66, 313–326.

Janssen, N.A.H., Lanki, T., Hoek, G., Vallius, M., de Hartog, J.J., Van Grieken, R., 2005. Associations between ambient, personal and indoor exposure to fine particulate matter constituents in Dutch and Finnish panels of cardiovascular patients. Occupational and Environmental Medicine 62, 868–877.

Jerrett, M., Buzzelli, M., Burnett, R.T., DeLuca, P.F., 2005. Particulate air pollution, social confounders, and mortality in small areas of an industrial city. Social Science and Medicine 60, 2845–2863.

Jordanova, N.V., Jordanova, D.V., Veneva, L., Yorova, K., Petrovsky, E., 2003. Magnetic response of soils and vegetation to heavy metal pollution – a case study. Environmental Science and Technology 37, 4417–4424.

Kardel, F., Wuyts, K., Maher, B.A., Hansard, R., Samson, R., 2011. Leaf saturation isothermal remanent magnetization (SIRM) as a proxy for particulate matter monitoring: interspecies differences and in-season variation. Atmospheric Environment 45, 5164–5171.

Kletetschka, G., Zila, V., Wasilewski, P.J., 2003. Magnetic anomalies on the tree trunks. Studia Geophysica et Geodaetica 47, 371–379.

Laltlanchhuanga, S.K., 2006. Studies of the Impact of Disturbance on Secondary Productivity of Forest Ecosystem with Special Reference to Surface, Sub-surface Litter Insect and Other Non-insect Groups (M.Sc. dissertation). Mizoram University.

Le Tertre, A., Medina, S., Samoli, E., Forsberg, B., Michelozzi, P., Boumghar, A., 2002. Short-term effects of particulate air pollution on cardiovascular diseases in eight European cities. Journal of Epidemiology and Community Health 56, 773–779.

Maher, B.A., 1998. Magnetic properties of modern soils and Quaternary loessic paleosols: paleoclimatic implications. Palaeogeography, Palaeoclimatology, Palaeoecology 137, 25–54.

Maher, B.A., 2009. Rain and dust: magnetic records of climate and pollution. Elements 5, 229–234.

Maher, B.A., Mooreb, C., Matzka, J., 2008. Spatial variation in vehicle-derived metal pollution identified by magnetic and elemental analysis of roadside tree leaves. Atmospheric Environment 42, 364–373.

Matzka, J., Maher, B.A., 1999. Magnetic biomonitoring of roadside tree leaves: identification of spatial and temporal variations in vehicle-derived particulates. Atmospheric Environment 33, 4565–4569.

Mitchell, R., Maher, B.A., Kinnersley, R., 2010. Rates of particulate pollution deposition onto leaf surfaces: temporal and interspecies magnetic analyses. Environmental Pollution 158 (5), 1472–1478.

Morawska, L., Zhang, J., 2002. Combustion sources of particles. 1. Health relevance and source signatures. Chemosphere 49, 1045–1058.

Moreno, E., Sagnotti, L., Dinarès-Turell, J., Winkler, A., Cascella, A., 2003. Biomonitoring of traffic air pollution in Rome using magnetic properties of tree leaves. Atmospheric Environment 37, 2967–2977.

Pandey, S.K., Tripathi, B.D., Prajapati, S.K., Mishra, V.K., Upadhyay, A.R., Rai, P.K., Sharma, A.P., 2005. Magnetic properties of vehicle-derived particulates and amelioration by Ficus infectoria: a keystone species. AMBIO: A Journal of the Human Environment 34 (8), 645–647.

Power, A.L., Worsley, A.T., Booth, C., 2009. Magneto-biomonitoring of intra-urban spatial variations of particulate matter using tree leaves. Environmental Geochemistry and Health 31, 315–325.

Prajapati, S.K., Pandey, S.K., Tripathi, B.D., 2006. Magnetic biomonitoring of roadside tree leaves as a proxy of vehicular pollution. Environmental Monitoring and Assessment 120, 169–175.

Rai, P.K., 2009. Comparative assessment of soil properties after bamboo flowering and death in a tropical forest of Indo-Burma hot spot. AMBIO: A Journal of the Human Environment 38 (2), 118–120.

Rai, P.K., 2011a. Dust deposition capacity of certain roadside plants in Aizawl, Mizoram: implications for environmental geomagnetic studies. In: Dwivedi, S.B., et al. (Ed.), Recent Advances in Civil Engineering, pp. 66–73.

Rai, P.K., 2011b. Biomonitoring of particulates through magnetic properties of road-side plant leaves. In: Tiwari, D. (Ed.), Advances in Environmental Chemistry. Excel India Publishers, New Delhi, pp. 34–37.

Rai, P.K., 2012. Assessment of multifaceted environmental issues and model development of an Indo-Burma hot spot region. Environmental Monitoring and Assessment 184, 113–131.

Rai, P.K., 2013. Environmental magnetic studies of particulates with special reference to biomagnetic monitoring using roadside plant leaves. Atmospheric Environment 72, 113–129.

Rai, P.K., Panda, L.S., 2014. Dust capturing potential and air pollution tolerance index (APTI) of some road side tree vegetation in Aizawl, Mizoram, India: an Indo-Burma hot spot region. Air Quality, Atmosphere and Health 7 (1), 93–101.

Rai, P.K., Panda, L.S., Chutia, B.M., Singh, M.M., 2013. Comparative assessment of air pollution tolerance index (APTI) in the industrial (Rourkela) and non industrial area (Aizawl) of India: an eco-management approach. African Journal of Environmental Science and Technology 7 (10), 944–948.

Sant'Ovaia, H., Lacerda, M.J., Gomes, C., 2012. Particle pollution – an environmental magnetism study using biocollectors located in northern Portugal. Atmospheric Environment 61, 340–349.

Sharma, A.P., Rai, P.K., Tripathi, B.D., 2007. Magnetic biomonitoring of roadside tree leaves as a proxy of vehicular pollution. In: Lakshmi, V. (Ed.), Urban Planning and Environment: Strategies and Challenges Macmillan Advanced Research Series, pp. 326–331.

Sharma, K., Singh, R., Barman, S.C., Mishra, D., Kumar, R., Negi, M.P.S., Mandal, S.K., Kisku, G.C., Khan, A.H., Kidwai, M.M., Bhargava, S.K., 2006. Comparison of trace metals concentration in PM_{10} of different location of Lucknow city, India. Bulletin of Environmental Contamination and Toxicology 77, 419–426.

Shu, J., Dearing, J., Morse, A., Yu, L., Li, C., 2000. Magnetic properties of daily sampled total suspended particulates in Shanghai. Environmental Science and Technology 34, 2393–2400.

Urbat, M., Lehndorff, E., Schwark, L., 2004. Biomonitoring of air quality in the Cologne conurbation using pine needles as a passive sampler part 1: magnetic properties. Atmospheric Environment 38, 3781–3792.

Walden, J., 1999. Sample collection and preparation. In: Walden, J., Oldfield, F., Smith, J.P. (Eds.), Environmental Magnetism: A Practical Guide. Quaternary Research Association, Cambridge, England, pp. 26–34. Technical guide no. 6.

Xie, S., Dearing, J.A., Bloemendal, J., 2000. The organic matter content of street dust in Liverpool, UK, and its association with dust magnetic properties. Atmospheric Environment 34, 269–275.

Xie, S., Dearing, J.A., Boyle, J.F., Bloemendal, J., Morse, A.P., 2001. Association between magnetic properties and element concentrations of Liverpool street dust and its implications.
Journal of Applied Geophysics 48, 83–92.

Yin, G., Hu, S., Cao, L., Roesler, W., Appel, E., 2013. Magnetic properties of tree leaves and
their significance in atmospheric particle pollution in Linfen city, China. Chinese Geographical Science 23 (1), 59–72.

CHAPTER TEN

Biomagnetic Monitoring of Particulate Matter through an Invasive Plant, *Lantana camara*

10.1 INTRODUCTION

Plant invasion is the priority threat to global biodiversity and hence deleterious to both the ecology and economy of any nation. Invasive plants or weeds transmogrify the landscapes of urban forests and duly affect its phytosociology as well as diversity of native species in a complex, intricate manner (Rai, 2013b). Various hypotheses have been proposed to understand the basic mechanism of succession in order to devise a sustainable management strategy, however, no one has described it in its totality (Rai, 2013b, 2015; Rai and Chutia, 2015).

Lantana camara is an important weed of agro and forest ecosystems, where it forms dense thickets that livestock cannot penetrate (Rai, 2015). The leaves are toxic when ingested by most domestic livestock or native mammals, although toxicity varies greatly between strains (Goulson and Derwent, 2004). In Australia (Leigh and Briggs, 1992; Groves and Willis, 1999), like other countries, plant invasion has been associated with the extinction of several valuable endemic plant species like *L. camara* (Gooden et al., 2009; Rai and Chutia, 2015).

Lantana camara are the dominant invasive species in many parts of Rajasthan resulting from landscape modernization (Robbins, 2001). Rai (2009, 2012a, 2013b, 2015) ecologically investigated *L. camara*, *Mikania micrantha*, and *Ageratum conyzoides* in the forests of an Indo–Burma hot spot region. In aquatic ecosystems of India several invasive plants like *Eichhornia crassipes* have been reported (Rai, 2008, 2009, 2012a; Rai and Chutia, 2015).

Lantana camara in spite of being an invasive species has several advantages. Osunkoya and Perrett (2011) demonstrated that under *Lantana* infested soil, moisture, pH, Ca, total and organic C, and total N were significantly elevated, while sodium, chloride, copper, iron, sulfur, and manganese, many of which can be toxic to plant growth if present in excess levels, were present

Biomagnetic Monitoring of Particulate Matter
ISBN 978-0-12-805135-1
http://dx.doi.org/10.1016/B978-0-12-805135-1.00010-X

181

at lower levels in soils compared to soils lacking *L. camara*. Another benefit of *Lantana* is that it improves soil hydraulic properties to benefit the wheat crop in a rice–wheat cropping sequence (Bhushan and Sharma, 2005; Rai, 2013b). Further, it may act on other invasives, e.g., the growth of the aquatic weed *Eichhornia crassipes* and the alga *Microcystis aeruginosa* may be inhibited by fallen leaves of *L. camara* (Kong et al., 2006; Rai, 2013b). Fleshy-fruited invasive plants provide food that supports indigenous frugivore populations (Gosper and Vivian-Smith, 2006; Rai, 2013b) and Gosper and Vivian-Smith (2006) using *L. camara* as a target species suggested that using the fruit characteristics of the invasive plant may assist to select replacement indigenous plants that are functionally similar from the perspective of frugivores.

Pollination success in diverse habitats, e.g., in the case of *L. camara*, *Ligustrum robustum*, and *Mimosa pigra* through profuse nectar and prolonged flower production (Ghazoul, 2002) aid in their invasion success. As demonstrated in the case of *L. camara*, forest gap/canopy openness plays a major role in invasive spread, therefore canopy intactness may be one of the prime management strategies that are rather difficult to maintain (Totland et al., 2005; Rai, 2013b). Many invasive aquatic plants like *E. crassipes* and also the terrestrial shrub *L. Camara* are reported to be very good in heavy metal as well as particulate pollution phytoremediation (Rai, 2008, 2009, 2012a; Rai and Panda, 2014; Rai and Chutia, 2015). Thus, the utilization of invasive plants in pollution abatement phytotechnologies may assist in their sustainable management.

Unfortunately, urban ecosystems of ecologically sensitive regions like an Indo-Burma hot spot are under severe air pollution stress (Rai, 2012a). Air pollutants comprised of both particulate matter (PM) and gaseous pollutants may cause adverse health effects in humans, affect plant life, and impact the global environment by changing the atmosphere of the earth (Rai et al., 2014; Rai, 2015; Rai and Chutia, 2015). Air pollution emanating from PM is particularly deleterious as these particulates lead to various cardiopulmonary diseases through oxidative stress (Rai, 2013b, 2015; Rai and Chutia, 2015).

Pertaining to biomagnetic monitoring, it is worth mentioning that early research demonstrated biogenic ferrimagnets to be present in the organisms like termites (Maher, 1998) and bacteria (Fassbinder et al., 1990). However, it is now well established through much additional research that urban PM may also contain magnetic particles along with other air pollutants (Pandey et al., 2005; Maher, 2009; Rai, 2013a; Rai et al., 2014).

In the light of abovementioned, this present study attempts to investigate the biomagnetic monitoring potential of *L. camara* (Figure 10.2) through magnetic measurements of plant leaves and concomitantly correlate these to ambient air quality (SPM and RSPM).

10.2 METHODOLOGY

Mizoram is the site of particular ecological relevance as it falls under an Indo-Burma hot spot region (Figure 10.1). In Mizoram, land use change through shifting cultivation is very frequent, which may exacerbate the problem of biological invasions (Rai, 2012b). The phytosociological studies

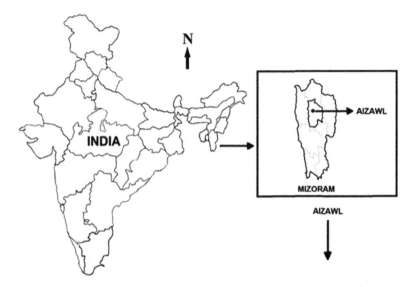

Figure 10.1 Map of the study area, Aizawl, Mizoram, northeast India.

Figure 10.2 Dust-loaded *Lantana camara* plant leaves in urban forests of Aizawl.

were performed at Aizawl, Mizoram, northeast India during the months of November to December 2013. It is worth noting that sites were selected in accordance with varying disturbance intensity. To perform phytosociological studies, quadrats of 10×10 m in size were randomly used. Quantitative/ phytosociological parameters such as % frequency, density, abundance, and total basal cover of each species present in quadrats were recorded and analyzed as per the methods of Kershaw (1973) and Misra (1968).

Pertaining to air quality parameters, the study was conducted seasonally, i.e., summer, rainy, and winter during 2013–2014.

10.2.1 Suspended Particulate Matter and Respirable Suspended Particulate Matter

Air pollutants such as suspended particulate matter (SPM) and respirable suspended particulate matter (RSPM) were analyzed for the four selected sites that were monitored by using a high volume air sampler (Envirotech model, APM-460NL) with gaseous attachment (Envirotech model, APM-411TE) monitoring 8 h per day during 2013–2014 with a frequency of twice in a season. The apparatus was kept at a height of 2 m from the surface of the ground. Once the sampling was over, the samples were brought to the laboratory and the concentrations of different pollutants were determined. RSPM were trapped by glass fiber filter papers (GF/A) of Whatman and SPM were collected in the separate containers at an average airflow rate of 1.5 m³/min.

As mentioned earlier, biomagnetic monitoring study was carried out in Aizawl district from four different sampling points (Figure 10.1): *Site 1.* Durtlang (urban area); *Site 2.* Zarkawt (urban area); *Site 3.* Ramrikawn (peri-urban area); *Site 4.* Mizoram University campus (MZU). MZU campus is an institutional area with low traffic density. Therefore, we selected MZU as reference or control site in order to compare the results recorded from other sites. Further, we used the winter season as dust or PM tends to concentrate during this season through atmospheric inversion (Verma and Singh, 2006) particularly during the morning hours. Further, in our recent research (Rai and Panda, 2014) we recorded maximum dust deposition during the winter season. This suggests that localized conditions like environmental, meteorological, or anthropogenic may be influencing or disturbing particulate deposition or it may reflect differences in the ability of leaf species to capture particulates (Rai et al., 2014).

The magnetic parameters such as magnetic susceptibility (χ), anhysteretic remanent magnetization (ARM), and saturation isothermal remanent magnetization (SIRM) were performed with dried leaves in 10-cc plastic sample pots at K.S. Krishnan Geomagnetic Research Lab of the Indian Institute of Geomagnetism, Allahabad, Uttar Pradesh, India.

The magnetic susceptibility indicates the total composition of the dust captured on the leaves, with a prevailing contribution from ferromagnetic minerals, which could show higher susceptibility values than paramagnetic and diamagnetic minerals, such as clay or quartz (Maher and Thompson, 1999; Evans and Heller, 2003; Sant'Ovaia et al., 2012). A Bartington (Oxford, England) MS-2B dual frequency susceptibility meter was used (Dearing, 1999) for measurements.

Anhysteretic remanent magnetization indicates the magnetic concentration and is also sensitive to the presence of fine grains ~0.04–1 µm (Thompson and Oldfield, 1986), which fall within the respirable size range of $PM_{2.5}$ and therefore have a high burden of toxicity (Power et al., 2009). The ARM was induced in samples using a Molspin (Newcastle-upon-Tyne, England) AF demagnetizer, whereby a direct current biasing field is generated in the presence of an alternating field, which peaks at 100 milli-Tesla (mT). The nature of this magnetic field magnetizes the fine magnetic grains, and the amount of magnetization retained within the sample (remanence) when removed from the field was measured using a Molspin1A magnetometer. The samples were then demagnetized to remove this induced field in preparation for the subsequent magnetic analysis (Walden, 1999).

SIRM reflects the total concentration of magnetic grains (Evans and Heller, 2003) and can be used as a proxy of PM concentration (Muxworthy et al., 2003). SIRM involves measuring the magnetic remanence of samples once they are removed from an induced field. Using a Molspin pulse magnetizer, an SIRM of 800 mT in the forward field was induced with the samples. At this high magnetization field, all magnetic grains within the sample become magnetized (Rai et al., 2014). The SIRM is actually the highest level of magnetic remanence that can be induced in a particular sample through application of high magnetic field; unit—Am^2. The instruments used for ARM and SIRM are fully automated.

The ratio of IRM-300 and SIRM is defined as the S-ratio (King and Channel, 1991), which mainly reflects the relative proportion of antiferromagnetic to ferrimagnetic minerals in a sample. A ratio close to 1.0 reflects almost pure magnetite while ratios of <0.8 indicate the presence of some antiferromagnetic minerals, generally goethite or hematite (Thompson, 1986).

10.2.2 Statistical Analysis

All statistical calculations were performed using Statistical Program for Social Science (SPSS version 11.2) and SAS software.

10.3 RESULTS AND DISCUSSION

Lantana camara was found to be the most dominant invasive weed as revealed through phytosociological studies (Table 10.1). The average seasonal values of two air pollutants (SPM and RSPM) recorded at four study sites throughout a one-year sampling period are presented in Table 10.2. The ambient PM concentrations were recorded highest at Ramrikawn, followed by Zarkawt and Durtlang, while the lowest values were recorded for the Tanhril area. The values of particulate pollutants were lowest in the rainy season, which may be because of large amounts of precipitation, whereas the summer and winter seasons were characterized by nearly the same concentration at all four different study sites. During the winter season there is increased atmospheric stability, which in turn allows for less general circulation and thus more stagnant air masses. It prevents an upward movement of air, hence atmospheric mixing is retarded and pollutants are trapped near the ground. Secondly, cold starts in winter lead to a longer period of incomplete combustion and longer warm-up times for catalytic converters, which generate more pollution (Shukla et al., 2010). Vehicular exhaust, construction work, commercial activities, the practice of *jhum* cultivation (slash and burn agriculture), and bad road conditions may be reasons for the augmented concentration of air pollutants at different study sites. In the present study, the quantity of RSPM and SPM at the four different sites were much higher than the prescribed limits of the Central Pollution Control Board of India during summer and winter season. The present concentrations of the air pollutants, specifically particulate pollutants, are high enough to affect human health of the area. Elevated levels of SPM and RSPM in air can cause respiratory diseases like asthma, cancer, blood pressure, etc. (Rai, 2015).

The average magnetic data collected throughout the one-year sampling period is presented in Table 10.3 for *L. camara* tree leaves. In Ramrikawn, the

Table 10.1 List of invasive weeds recorded at the four sites of Aizawl, Mizoram, Northeast India (an Indo-Burma hot spot region)

Name of species	Q1	Q2	Q3	Q4	Q5	No. of individuals	Density	Freq.	Abundance	Basal area	Basal cover	Relative density	Relative frequency	Relative abundance	IVI
Lantana camara	+	+	−	+	+	20	4	80	5	78.55	314.2	4.950495	8.510638	34.25957	47.7207
Ageratum conizoides	+	+	+	+	+	65	13	100	13	7.06	91.78	16.08911	10.6383	10.00746	36.73487
Spilanthes oleracea	+	+	+	+	+	70	14	100	14	3.14	43.96	17.32673	10.6383	4.793287	32.75832
Biden biternata	+	+	−	+	+	32	6.4	80	8	19.63	125.632	7.920792	8.510638	13.69859	30.13002
Spilanthes sp.	−	−	+	−	+	14	2.8	40	7	4.9	13.72	3.465347	4.255319	1.495994	9.21666
Mikania micrantha	+	+	+	+	+	31	6.2	100	6.2	7.06	43.772	7.673267	10.6383	4.772788	23.08436
Clerodendron infortunatum	+	+	−	−	−	8	1.6	40	4	78.55	125.68	1.980198	4.255319	13.70383	19.93935
Imperata cylindrica	+	+	+	+	+	42	8.4	100	8.4	4.9	41.16	10.39604	10.6383	4.487982	25.52232
Par eng (local name)	+	+	+	−	−	19	3.8	60	6.3	3.14	11.932	4.70297	6.382979	1.301035	12.38698
Unidentified															
Merremia umbellatum	+	+	+	−	+	22	4.4	80	5.5	19.63	86.372	5.445545	8.510638	9.417784	23.37397
Panicum conjugatum	+	+	+	−	+	32	6.4	80	8	1.76	11.264	7.920792	8.510638	1.228198	17.65963
Kyllingia brevifolia	+	−	+	+	+	49	9.8	80	12.2	0.78	7.644	12.12871	8.510638	0.833482	21.47283

Table 10.2 The average concentration of two air pollutants (SPM and RSPM) at different study sites during 2013–2014

Air pollutants	Ramrikawn			Tanhril			Zarkawt			Durtlang			CPCB standard (residential and rural area)
	Summer	Winter	Rainy	Summer	Winter	Rainy	Summer	Winter	Rainy	Summer	Winter	Rainy	
SPM (μg m^{-3})	263.12 ± 0.01	260.01 ± 0.12	98.04 ± 0.04	210.91 ± 0.16	207.07 ± 0.41	42.9 ± 0.21	223.51 ± 0.11	229.21 ± 0.02	93.01 ± 0.29	220.22 ± 0.24	224.07 ± 0.01	87.03 ± 0.32	200
RSPM (μg m^{-3})	228.09 ± 0.23	232.23 ± 0.19	71.21 ± 0.83	102.31 ± 0.02	109.28 ± 0.04	20.18 ± 0.12	189.03 ± 0.08	200.61 ± 0.41	63.18 ± 0.19	183.41 ± 0.03	190.15 ± 0.11	56.91 ± 0.05	100

SPM, suspended particulate matter; RSPM, respirable suspended particulate matter; CPCB, Central Pollution Control Board, New Delhi, India.

Table 10.3 Summary of the magnetic data (mean and standard deviation) for roadside dust on *Lantana camara* leaves in the different sampling sites

Sites	χ (10^{-7} m^3 kg^{-1})		ARM (10^{-5} Am^2 kg^{-1})		SIRM (10^{-5} Am^2 kg^{-1})		ARM/χ (10^2 Am^{-1})		SIRM/χ (10^2 Am^{-1})		S-ratio	
	2013–14 (Winter)	2013–14 (Summer)	2013–14 (Winter)	2013–14 (Summer)	2013–14 (Winter)	2013–14 (Summer)	2013–14 (Winter)	2013–14 (Summer)	2013–14 (Winter)	2013–14 (Summer)	2013–14 (Winter)	2013–14 (Summer)
Ramrikawn	37.09 ± 0.81	23.21 ± 0.08	8.24 ± 0.31	22.01 ± 0.17	203.70 ± 0.52	271.51 ± 0.29	0.22	0.94	5.49	11.69	0.951	0.944
Tanhril	19.72 ± 0.41	11.81 ± 0.07	7.19 ± 0.18	10.81 ± 0.17	140.41 ± 0.44	132.77 ± 0.05	0.36	0.91	7.12	11.24	0.954	0.867
Zarkawt	33.87 ± 0.54	20.75 ± 0.18	8.19 ± 0.41	20.05 ± 0.08	201.42 ± 0.26	244.31 ± 0.12	0.24	0.96	5.94	11.77	0.952	0.931
Durtlang	28.59 ± 0.39	14.03 ± 0.11	4.48 ± 0.29	12.56 ± 0.41	153.21 ± 0.31	153.42 ± 0.71	0.15	0.89	5.35	10.93	0.963	0.891

χ, ARM, and SIRM values of *L. camara* were 37.09 ± 0.81 $(10^{-7} \text{m}^3 \text{kg}^{-1})$, 8.24 ± 0.31 $(10^{-5} \text{Am}^2 \text{kg}^{-1})$, and 203.70 ± 0.52 $(10^{-5} \text{Am}^2 \text{kg}^{-1})$ for the winter season and 23.21 ± 0.08 $(10^{-7} \text{m}^3 \text{kg}^{-1})$, 22.01 ± 0.17 $(10^{-5} \text{Am}^2 \text{kg}^{-1})$, and 271.51 ± 0.29 $(10^{-5} \text{Am}^2 \text{kg}^{-1})$, respectively, for the summer season. In Tanhril, the χ, ARM, and SIRM values of *L. camara* were 19.72 ± 0.41 $(10^{-7} \text{m}^3 \text{kg}^{-1})$, 7.19 ± 0.18 $(10^{-5} \text{Am}^2 \text{kg}^{-1})$, and 140.41 ± 0.44 $(10^{-5} \text{Am}^2 \text{kg}^{-1})$ for the winter season and 11.81 ± 0.07 $(10^{-7} \text{m}^3 \text{kg}^{-1})$, 10.81 ± 0.17 $(10^{-5} \text{Am}^2 \text{kg}^{-1})$, and 132.77 ± 0.05 $(10^{-5} \text{Am}^2 \text{kg}^{-1})$, respectively, for the summer season. In Zarkawt, the χ, ARM, and SIRM values of *L. camara* were 33.87 ± 0.54 $(10^{-7} \text{m}^3 \text{kg}^{-1})$, 8.19 ± 0.41 $(10^{-5} \text{Am}^2 \text{kg}^{-1})$, and 201.42 ± 0.26 $(10^{-5} \text{Am}^2 \text{kg}^{-1})$ for the winter season and 20.75 ± 0.18 $(10^{-7} \text{m}^3 \text{kg}^{-1})$, 20.05 ± 0.08 $(10^{-5} \text{Am}^2 \text{kg}^{-1})$, and 244.31 ± 0.12 $(10^{-5} \text{Am}^2 \text{kg}^{-1})$, respectively, for the summer season. In Durtlang, the χ, ARM, and SIRM values of *L. camara* were 28.59 ± 0.39 $(10^{-7} \text{m}^3 \text{kg}^{-1})$, 4.48 ± 0.29 $(10^{-5} \text{Am}^2 \text{kg}^{-1})$, and 153.21 ± 0.31 $(10^{-5} \text{Am}^2 \text{kg}^{-1})$ for the winter season and 14.03 ± 0.11 $(10^{-7} \text{m}^3 \text{kg}^{-1})$, 12.56 ± 0.41 $(10^{-5} \text{Am}^2 \text{kg}^{-1})$, and 153.42 ± 0.71 $(10^{-5} \text{Am}^2 \text{kg}^{-1})$, respectively, for the summer season.

The high dispersion degrees of susceptibility and remanent magnetism mainly result from the sampling sites in different functional areas. Samples collected in the rural area (Tanhril) show low susceptibility and remanent magnetism, whereas tree leaves sampled in the city (Zarkawt and Durtlang) and peri-urban (Ramrikawn) area show higher values. The values of ARM/χ and SIRM/χ can reflect the grain size of magnetic minerals (Evans and Heller, 2003). The results show that the ARM/χ and SIRM/χ values are found to be low at all studied sites (Table 10.3). ARM/χ values ranged from 0.15 to 0.96 (10^2Am^{-1}) and SIRM/χ values range from 5.35 to 11.77 (10^2Am^{-1}), respectively for all study sites. Low values of ARM/χ and SIRM/χ indicate relatively large grain size of magnetic particles present in leaf samples (Yin et al., 2013). The S-ratio of *L. camara* leaf samples ranges from 0.867 to 0.963 (Table 10.3) for the four different study sites, which means that these leaf samples are dominated by "soft" magnetic minerals with a low coercive force, but a minor part of "hard" magnetic minerals with a relatively high coercive force also exists (Robinson, 1986).

From the findings recorded in Table 10.3, we can conclude that Ramrikawn site shows slightly higher magnetic values compared to the other sites. On the other hand, Ramrikawn and Zarkawt experience relatively higher deposition of magnetic grains, originating from PM. The spatial trends of these three magnetic parameters display similar

Table 10.4 Correlation between magnetic measurements of *Lantana camara* with SPM and RSPM at different study sites during 2013–2014

Magnetic parameter	SPM (R^2)		RSPM (R^2)	
	Winter	Summer	Winter	Summer
χ	0.815	0.688	0.955	0.733
ARM	0.160	0.646	0.032	0.688
SIRM	0.665	0.663	0.695	0.689

trends of having Ramrikawn at the maximum value and Tanhril area at the lowest value. The correlation coefficients indicate a significant relationship between the concentration of PM and magnetic measurement for *L. camara* tree leaves (Table 10.4). Hansard et al. (2011) studied atmospheric particle pollution emitted by a combustion plant using the tree leaves. Results show that a significant correlation is obtained between the SIRM and PM_{10}. Hu et al. (2008) also observed a good correlation of magnetic parameters (χ, ARM, and SIRM) with air pollutants, particularly heavy metals. Further, Kardel et al. (2011) recorded a significant correlation between leaf SIRM and ambient PM concentrations. The other studies also demonstrated a significant correlation between magnetic parameter and PM as studied elsewhere (Pandey et al., 2005; Prajapati et al., 2006). Muxworthy et al. (2003) advocated that the value of SIRM is strongly correlated with the PM mass. This not only acts as a proxy for PM monitoring but also is a viable alternative to magnetic susceptibility since the samples are magnetically too weak.

The average magnetic concentration data (Table 10.3) demonstrate that the accumulation of PM on tree leaves varies across the four different studied locations. The results suggest that Ramrikawn and Zarkawt experience the heaviest load of particulates in comparison to the low-deposition sites of Durtlang and Tanhril area. Ramrikawn recorded the highest values of magnetic parameters, which may be attributed to heavy vehicular load (due to presence of FCI, India), street dust, and dust from fragile rocks. Zarkawt and Durtlang may have vehicular pollution as the only source of PM, while Tanhril, being a village area, is relatively free from vehicular pollution and other anthropogenic activities.

The processes that are responsible for large particulate deposition on leaves are sedimentation under gravity, diffusion, and turbulent transfer giving rise to impaction and interception (Speak et al., 2012; Rai and Chutia, 2015). Mitchell et al. (2010) emphasized complex dependence of deposition velocities (*v*d) on different variables such as particle size and density, terrain vegetation, and chemical species. Further, landscape geography and architecture may also affect particulate concentration and its deposition on vegetation. Also, the dust collection capacity of plants depends on the shape and surface geometry of plant leaves, leaf size, and characteristics such as roughness, porosity, plant height, canopy and aspect, and distance from emission road and buildings (Rai and Chutia, 2015).

10.4 CONCLUSIONS

Lantana camara is a troublesome invasive weed in the Indian subcontinent and other countries. However, as a good PM accumulator it may sustainably ameliorate the environment burdened with PM pollution.

REFERENCES

Bhushan, L., Sharma, P.K., 2005. Long-term effects of lantana residue additions on water retention and transmission properties of a medium-textured soil under rice–wheat cropping in northwest India. Soil Use and Management 21, 32–37.

Dearing, J.A., 1999. Environmental Magnetic Susceptibility: Using the Bartington MS2 System. British Library Cataloguing in Publication Data.

Evans, M.E., Heller, F., 2003. Environmental Magnetism: Principles and Applications of Enviromagnetics (International Geophysics). Academic Press. Elsevier, London. 299 pp.

Fassbinder, J.W.E., Stanjek, H., Vali, H., 1990. Occurrence of magnetic bacteria in soil. Nature 34, 161–163.

Ghazoul, J., 2002. Flowers at the front line of invasion? Ecological Entomology 27, 638–640.

Gooden, B., French, K., Turner, P.J., 2009. Invasion and management of a woody plant, *Lantana camara* L., alters vegetation diversity within wet sclerophyll forest in southeastern Australia. Forest Ecology and Management 257, 960–967.

Gosper, C.R., Vivian Smith, G., 2006. Selecting replacements for invasive plants to support frugivores in highly modified sites: a case study focusing on *Lantana camara*. Ecological Management and Restoration 7 (3), 197–203.

Goulson, D., Derwent, L.C., 2004. Synergistic interactions between an exotic honeybee and an exotic weed: pollination of *Lantana camara* in Australia. Weed Research 44, 195–202.

Groves, R.H., Willis, A.J., 1999. Environmental weeds and loss of native plant biodiversity: some Australian examples. Australian Journal of Environmental Management 6, 164–171.

Hansard, R., Maher, B.A., Kinnersley, R., 2011. Biomagnetic monitoring of industry-derived particulate pollution. Environmental Pollution 159 (6), 1673–1681.

Hu, S.Y., Duan, X.M., Shen, M.J., et al., 2008. Magnetic response to atmospheric heavy metal pollution recorded by dust-loaded leaves in Shougang industrial area, western Beijing. Chinese Science Bulletin 53 (10), 1555–1564.

Kardel, F., Wuyts, K., Maher, B.A., Hansard, R., Samson, R., 2011. Leaf saturation isothermal remanent magnetization (SIRM) as a proxy for particulate matter monitoring: interspecies differences and in-season variation. Atmospheric Environment 45, 5164–5171.

Kershaw, R.A., 1973. Quantitative and Dynamic Plant Ecology. Edward Arnold Ltd, London.

King, J.W., Channel, J.E.T., 1991. Sedimentary magnetism, environmental magnetism and magnetostratigraphy. Reviews of Geophysics 29, 358–370.

Kong, C.H., Wang, P., Zhang, C.X., Zhang, M.X., Hu, F., 2006. Herbicidal potential of allelochemicals from *Lantana camara* against *Eichhornia crassipes* and the alga *Microcystis aeruginosa*. European Weed Research Society 46, 290–295.

Leigh, J.H., Briggs, J.D., 1992. Threatened Australian Plants: Overview and Case Studies. Australian National Parks and Wildlife Service, Canberra.

Maher, B.A., 1998. Magnetic biomineralisation in termites. Proceedings of the Royal Society of London 265, 233–237.

Maher, B.A., 2009. Rain and dust: magnetic records of climate and pollution. Elements 5, 229–234.

Maher, B.A., Thompson, R., 1999. Quaternary, Climates, Environments and Magnetism. Cambride University Press. 390 pp.

Mishra, R., 1968. Ecology Work Book. Oxford and IBH Publishing Co., New Delhi.

Mitchell, R., Maher, B.A., Kinnersley, R., 2010. Rates of particulate pollution deposition onto leaf surfaces: temporal and interspecies magnetic analyses. Environmental Pollution 158 (5), 1472–1478.

Muxworthy, A.R., Matzka, J., Davila, A.F., Petersen, N., 2003. Magnetic signature of daily sampled urban atmospheric particles. Atmospheric Environment 37, 4163–4169.

Osunkoya, O.O., Perrett, C., 2011. *Lantana camara* L. (Verbenaceae) invasion effects on soil physicochemical properties. Biology and Fertility of Soils 47, 349–355.

Pandey, S.K., Tripathi, B.D., Prajapati, S.K., Mishra, V.K., Upadhyay, A.R., Rai, P.K., Sharma, A.P., 2005. Magnetic properties of vehicle derived particulates and amelioration by *Ficus infectoria*: a keystone species. AMBIO: A Journal on Human Environment 34 (8), 645–647.

Power, A.L., Worsley, A.T., Booth, C., 2009. Magneto-biomonitoring of intraurban spatial variations of particulate matter using tree leaves. Environmental Geochemistry and Health 31, 315–325.

Prajapati, S.K., Pandey, S.K., Tripathi, B.D., 2006. Magnetic biomonitoring of roadside tree leaves as a proxy of vehicular pollution. Environmental Monitoring and Assessment 120, 169–175.

Rai, P.K., 2013a. Environmental magnetic studies of particulates with special reference to bio magnetic monitoring using roadside plant leaves. Atmospheric Environment 72, 113–129.

Rai, P.K., 2013b. Plant Invasion Ecology: Impacts and Sustainable Management. Nova Science Publisher, New York, p. 196.

Rai, P.K., 2008. Heavy-metal pollution in aquatic ecosystems and its phytoremediation using wetland plants: an eco-sustainable approach. International Journal of Phytoremediation 10 (2), 133–160.

Rai, P.K., 2009. Heavy metal phytoremediation from aquatic ecosystems with special reference to macrophytes. Critical Reviews in Environmental Science and Technology 39 (9), 697–753.

Rai, P.K., 2012a. An eco-sustainable green approach for heavy metals management: two case studies of developing industrial region. Environmental Monitoring and Assessment 184, 421–448.

Rai, P.K., 2012b. Assessment of multifaceted environmental issues and model development of an Indo-Burma hot spot region. Environmental Monitoring and Assessment 184, 113–131.

Rai, P.K., 2015. Concept of plant invasion ecology as prime factor for biodiversity crisis: introductory review. International Research Journal of Environmental Sciences 4 (5), 85–90.

Rai, P.K., Panda, L.S., 2014. Dust capturing potential and air pollution tolerance index (APTI) of some roadside tree vegetation in Aizawl, Mizoram, India: an Indo-Burma hot spot region, air quality. Atmosphere and Health 7 (1), 93–101.

Rai, P.K., Chutia, B., 2015. Biomagnetic monitoring of particulate matter (PM) through leaves of an invasive alien plant *Lantana camara* in an Indo-Burma hot spot region. Environmental Technology and Innovation (Submitted to Journal).

Rai, P.K., Chutia, B.M., Patil, S.K., 2014. Monitoring of spatial variations of particulate matter (PM) pollution through bio-magnetic aspects of roadside plant leaves in an Indo-Burma hot spot region. Urban Forestry and Urban Greening 13, 761–770.

Robbins, P., 2001. Tracking invasive land covers in India or why our landscapes have never been modern. Annals of the Association of American Geographers 91, 637–659.

Robinson, S.G., 1986. The late Pleistocene palaeoclimatic record of North Atlantic deep-sea sediments revealed by mineral-magnetic measurements. Physics of the Earth and Planetary Interiors 42 (1–2), 22–47.

Sant'Ovaia, H., Lacerda, M.J., Gomes, C., 2012. Particle pollution—an environmental magnetism study using biocollectors located in northern Portugal. Atmospheric Environment 61, 340–349.

Shukla, V., Dalal, P., Chaudhry, D., 2010. Impact of vehicular exhaust on ambient air quality of Rohtak city, India. Journal of Environmental Biology 31 (6), 929–932.

Speak, A.F., Rothwell, J.J., Lindley, S.J., Smith, C.L., 2012. Urban particulate pollution reduction by four species of green roof vegetation in a UK city. Atmospheric Environment 61, 283–293.

Thompson, R., 1986. Modelling magnetization data using SIMPLEX. Physics of the Earth Planetary Interiors 42, 113–127.

Thompson, R., Oldfield, F., 1986. Environmental magnetism. Allen and Unwin, London, 27 pp. trunks. Studia Geophysica et Geodaetica 47, 371–379.

Totland, O., Nyeko, P., Bjerknes, A., Hegland, S.J., Nielsen, A., 2005. Does forest gap size affects population size, plant size, reproductive success and pollinator visitation in *Lantana camara*, a tropical invasive shrub? Forest Ecology and Management 215, 329–338.

Verma, A., Singh, S.N., 2006. Biochemical and ultra-structural changes in plant foliage exposed to auto-pollution. Environmental Monitoring and Assessment 120, 585–602.

Walden, J., 1999. Sample collection and preparation. In: Walden, J., Oldfield, F., Smith, J.P. (Eds.), Environmental Magnetism: A Practical Guide. Quaternary Research Association, Cambridge, England, pp. 26–34. Technical guide no. 6.

Yin, G., Hu, S., Cao, L., Roesler, W., Appel, E., 2013. Magnetic properties of tree leaves and their significance in atmospheric particle pollution in Linfen city, China. Chinese Geographical Science 23 (1), 59–72.

INDEX

Printed in the United States
By Bookmasters